Biosphere and Environmental Safety

V. I. Osipov

Biosphere and Environmental Safety

 Springer

V. I. Osipov
Institute of Environmental Geoscience,
    Russian Academy of Sciences
Moscow
Russia

ISBN 978-3-030-08209-3   ISBN 978-3-319-91259-2   (eBook)
https://doi.org/10.1007/978-3-319-91259-2

Preface

Ensuring ecological security is among the global strategic tasks. This problem is related to the rapidly aggravating threats caused by increasing population growth on the Earth, diminishing life-supporting resources, technogenesis, global climate change, and escalating natural disasters. Along with military security, ecological security becomes the most important factor controlling human survival on the Earth.

This book examines the state of natural environment and the causes of its degradation using the biospheric approach. The urgent global tasks are identified following the ideas first expressed by V. I. Vernadsky and aimed at the search for the new principles of human–nature interaction.

In 2017, the book "Biosphere and Environmental Security" was published in Russian. The present edition is based on the English translation of the original manuscript with the minor improvement and corrections introduced. The author gratefully acknowledges Tatyana P. Kolchugina for the highly professional English translation of the manuscript and valuable comments and suggestions, as well as Olga N. Eremina for her assistance in the editorial preparation of the manuscript.

Moscow, Russia   V. I. Osipov

Contents

Biosphere and Environmental Safety   1
1   Introduction   1
2   The Biosphere Is the Sphere of Life Not Yet Completely Understood .   2
3   How Many of Us Will There Be? Earth's Population   6
4   Technogenesis and Its Influence on the State
      of the Biosphere   9
      4.1   Technogenesis Is the Most Important Factor in the Destruction
            of the Natural Environment   9
      4.2   The Code of Human Technogenic Activity   12
5   "Population Load." The Human Adaptation Corridor   13
      5.1   Technogenesis and Human "Population Load".   13
      5.2   Adaptation Corridor   14
6   What Happens to Climate?   15
      6.1   Change in Natural Climate Systems in the History
            of Earth   15
      6.2   Causes of Climate Change   17
      6.3   The Role of Technogenesis in the Global Temperature Change
            on Earth   19
7   Life and Natural Disasters Risk   22
      7.1   Intensification of Natural Disasters   22
      7.2   Internal Geodynamics of Earth (Seismic Phenomena).   23
      7.3   Dangerous Technogenic Processes   26
      7.4   The Effects of Global Climate Change   27
      7.5   Socio-economic Losses   32
      7.6   Assessment of Natural Risks   33
8   Market Economy and Ecology   35
9   Protection of Nature and Nature Management. Environmental
      Security   38
      9.1   Conservation of Nature   38

  9.2   Nature Management   39
  9.3   Environmental Security   41
10   Nature as a Living Soul. Ethical Aspect   42
  10.1   Life with Nature   42
  10.2   A Live Drop of Water   43
  10.3   Green Economy   43
  10.4   Ethical Principles and Nature   44
11   The Future of Civilization. V. I. Vernadsky
     and the Noosphere   45
  11.1   The History of Environmental Crises on Earth   45
  11.2   A Brewing Ecological Crisis   46
  11.3   Will Civilization Survive?   47
12   Conclusion   49
References   51

About the Author

Prof. V. I. Osipov is Dr. Sci. (Geol.-Min.), Full Member of the Russian Academy of Sciences, and Honorary Professor of M. V. Lomonosov Moscow State University. He is Laureate of the State Prize of the USSR and the State Award of the Russian Government in Science and Technology, Editor in Chief of "Geoekologiya" ("Environmental geoscience") journal, and Scientific Adviser of Sergeev Institute of Environmental Geoscience RAS. The main areas of his scientific research are physicochemical mechanics of soils and rocks, natural and technonatural hazardous processes, assessment of natural risks, and geoecological problems. He is the author of more than 650 scientific publications, including 22 monographs and textbooks, and 23 patents.

Biosphere and Environmental Safety

1 Introduction

The term "ecology" was introduced into scientific lexicon by the German biologist Ernest Haeckel (Greek οἶκος—"habitat," "house"; λογία—"study of"). The concept of ecology resonates with the notion of economics: ecology is the science of the surrounding environment or "a house," and the economy, in translation from Greek, means the art of housekeeping. The common house for mankind is its habitat—the biosphere. The biosphere provides free life-sustaining resources to humans. These resources gradually disappear with the degradation of the biosphere and, thereby, the potential for the continued human existence on Earth is diminishing.

The content of ecology is constantly expanding; it is becoming an interdisciplinary field of science. Today, this term represents the doctrine of the biosphere and encompasses human habitat, the features of its development, and the role of civilization in this process.

The problems of ecology are now acquiring the most important strategic importance since they are about the most crucial thing—the lasting existence of human civilization on Earth.

Growth of energy-and-technology capabilities has led to a manyfold increase in the consumption of natural resources and the degree of the impact on the biosphere. The civilization's might has become an unavoidable factor that can disrupt the whole system of life on Earth and lead to unpredictable consequences.

The biosphere went beyond its ecological capacity on large part of Earth under the influence of humans as early as at the beginning of the 20th century, which disrupted the stability of the environment and the conditions necessary for the normal existence of all living beings. This process intensified particularly in the second half of the 20th and the beginning of the 21th centuries and included global climate change, progressive pollution of the environment, development of natural catastrophic phenomena, shortage of drinking water, disruption of human health, and destruction and depletion of soil cover.

Unfortunately, human society still does not completely perceive the complexity of the modern era and major changes that are the consequences of the global process of degradation of the natural environment. Despite the critical processes developing in the biosphere, scientific and technological research is still directed toward the support of intense and massive use of biospheric natural resources to increase economic growth, prosperity, and consumption, with the ultimate goal of creating the "paradise on Earth."

The euphoria of economic success and the introduction of modern scientific achievements that expand the impact on the biosphere do not yet allow us to develop rigorous research on the real situation in the biosphere and to identify the laws of its development and the extreme (threshold) parameters of technogenesis. This is now the main problem of ecology and the main challenge to humankind.

The biosphere keeps many secrets that are yet to be revealed. Modern ideas about the development of the biosphere suggest the inevitability of the transition from tranquility of the Darwinian development to periodic catastrophic restructuring. The evolution of the biosphere is a chain of catastrophes with unpredictable outcomes. Henri Poincaré called such a state "bifurcational."

Without understanding of the laws governing the development of the biosphere, even at a conceptual level, it is difficult to perceive its evolution and the conditions for the existence of human society, not only in the long-term, but also even in the short-term context. To know this process, one needs to comprehend the strategy of the planetary consciousness. Therefore, all the arguments about sustainable development, the "golden billion," or Earth Charter, which are so-in now, without a serious scientific justification remain just a figment of fantasy and intentions to treat desired as real. Besides, some environmental doctrines have underlying political or economic reasons.

Below, we discuss environmental changes in the era of modern technogenesis. We analyze problems of degradation of terrestrial ecosystems and justify the need for profound scientific research of the laws of the evolution of the biosphere as the foundation for the global consciousness. This would allow us to forecast better the future of human civilization. Solution to these problems requires establishing fundamental research concepts and developing non-trivial ideas that can radically change our understanding of the possible paths for human survival.

2   The Biosphere Is the Sphere of Life Not Yet Completely Understood

The term *biosphere* encompasses all the structures and functions of living, inert, and bio-inert systems that together constitute life in the form realized on the planet Earth. It is generally accepted that Earth's biosphere is the universal planetary and human asset and the main productive force.

Living organisms (animals and plants) of the biosphere represent *biota*. They have three major functions: (1) destructive—the ability to decompose substances, accumulate chemical elements, and involve them in new geochemical cycles, (2) energy—the ability to absorb and transform energy; and (3) information—the ability to transfer genetic information.

Biota consists of *biocenoses*, i.e., elementary communities of organisms. In biogeocoenoses, the integrity and interdependence of biota (*biocenosis*) and its environment (*biotope*) is maintained (Sukachev 1964).

Inert systems are formed from the abiotic components of the biosphere—the atmosphere, water, and rocks. The inert environment includes minerals, rocks, water, and air. The inert systems are the environment where biota develops and can be solid, liquid, or gaseous bodies formed on Earth or in space. The inert and living objects are being constantly transformed, giving rise to new geochemical cycles of elements. Therefore, all the equilibria formed in the biosphere are transient and dynamic.

Inert matter lacks genetic unity and the ability to reproduce: it is formed but does not propagate. Intrinsic forces of evolution are not inherent in inert substances. The formation and change of inert bodies is influenced by physicochemical factors: changes in temperature, pressure, and chemical composition of the external environment. Processes that create inert matter are reversible, and the processes underlying the formation of living matter are irreversible. Therefore, from a point of view about genesis of biota, it is possible to apply the following statement: "everything alive is from the living" and "everything inert is from physical and chemical conditions" (Naumov 2010).

Both living and inert substances have molecular structure, and this feature unites them. However, they are fundamentally different: living organisms exist only in the biosphere in the form of discrete bodies that can reproduce, which determines the genetic connections between them. The ability of living organisms to reproduce depends on the availability of life-sustaining products—chemical substances and energy. If unlimited sources of life-sustaining products are available, the reproduction of living organisms can proceed at a catastrophic rate. For example, it is estimated that under favorable conditions, a diatom in eight days can produce similar organisms in a volume close to the size of our planet, and then, in an hour, double it.

Inert matter and living organisms compose bio-inert systems of the biosphere. The most favorable conditions for the formation of bio-inert systems exist at the boundaries between the lithosphere, atmosphere, and hydrosphere. Air, surface- and groundwater, soils, and rocks are saturated with microorganisms—living organisms in bio-inert systems. They are especially abundant in silts and soils, reaching several billion per gram.

The role of microorganisms in bio-inert systems is extremely high; they control many processes, from accumulation of solar energy on Earth to the state of human health (both good and poor). About 10,000 species of microorganisms live in the human body, including bacteria, fungi, and viruses. A person would not be able to survive if all microorganisms are removed from the body. One of the examples of negative effects on the human body is the activity of "stone-forming" nanobacteria discovered by an American geologist Robert Folk (Folk 1998) who described the phenomenon of human body biomineralization associated with these tiny bacteria, fractions of micron in size, that form hard mineral structures (kidney stones, atheromatous plaques of blood vessels, etc.) in the human body, which affects the functioning of human organs (Volkov et al. 2004).

Live, inert, and bio-inert systems compose the biosphere of the planet. Elementary structural and functional units of the biosphere and their complexes are called ecosystems (Sukachev 1964).

The biosphere—the shell of Earth, connecting the atmosphere, hydrosphere, and part of the solid Earth, is a reservoir of living matter. Living matter by weight does not exceed 0.25% of the entire biosphere. It is believed that the boundaries of the biosphere are determined by the spread of certain species of bacteria in the atmosphere and Earth's crust. At present, bacteria are found in the atmosphere within the ozone layer at a distance of 25–30 km from the surface of Earth. Thus, the upper limit of the biosphere is limited to the ozone layer. The lower boundary of the biosphere is determined by the thickness of the hydrosphere: on the continents, it goes to a depth of 4–5 km, and in the oceans, it is 0.5–1.0 km down from the ocean floor.

The existence of the biosphere is inseparably linked with the history of Earth as a cosmic body. The biosphere has been transforming through the geological history of Earth and its evolution continues to the present day. We can only get the most general idea of this transformation over more than four billion years of the biosphere's history through indirect paleontological and historical-geological data, and only for the last few hundred million years. The main landmarks of this transformation are the emergence of the atmosphere and the hydrosphere, the first signs of life in the form of simple marine and amphibious living organisms, the emergence of Earth's green cover and mammals, and, finally, the rise of "*Homo sapiens*" (Budyko 1984).

Examination of the structure of the biosphere and its functioning allows formulating the most important philosophical concept: what is life? The Russian scientist and Academician V. I. Vernadsky was essentially the first to answer this question, "*Life is a phenomenon of animate and inanimate nature, representing the manifestation of a single process from the planetary point of view.*" Life means the existence of complex and organized by evolution bio-inert systems. In 1944, V. I. Vernadsky formulated four postulates that are important for life on the planet and the existence of humankind in the context of the biosphere: (1) the biosphere has its own laws, different from the laws of life of human society; (2) biota (living matter) controls the entire biosphere and the interactions in it, i.e., determines biological regulation in the biosphere; (3) biota created the environment and together with it forms the biosphere; and (4) humans must study and observe the laws of the biosphere.

Living matter generates life-sustaining resources on Earth: it is the main accumulator of cosmic (solar) energy entering the biosphere stored in coal, oil, combustible gas, and dispersed organic matter of soils.

The biosphere has two main sources of energy: endogenous, coming from the depths of Earth, and exogenous—cosmic. The dispute between the supporters of these two positions on the sources of energy goes deep into history and is known as the "plutonist"–"neptunist" controversy. Endogenous energy flow consists of energy associated with radioactive decay, gravitation, tidal friction, and tectonic and meta-morphic processes. The average total flux of endogenous energy is about 10 J/(m²s).

The flow of energy coming to Earth from the Sun is three orders of magnitude greater than energy flow from the depths of Earth. However, most part of solar energy is reflected back into space. The amount of solar energy assimilated by green matter of the planet is 40–60 J/(m²s). This value is comparable to the amount of energy entering the biosphere from the deep layers of Earth.

The energy transmitted by the Sun to Earth does not disrupt the balance of energy exchange between them; however, the exchanged energy differs: Earth receives short-wave radiation from the Sun and emits energy in the infrared spectrum. This enables physical and chemical processes promoting the evolution of the biosphere (Moiseyev 2007).

The modern biosphere is the most complicated unique structure of internal (intra-biospheric) and external (inter-cosmic) interactions with an enormous number of internal connections and processes, from molecular to geodynamic of entire conti-nents. Goethe wrote, *"Nature is the only book with a great content on each sheet."*

The biosphere is the cradle and home of humanity. It has properties not yet fully comprehensible to humans. The flow of information in the biosphere exceeds by orders of magnitude the flow of information in all computers of the world. Accord-ing to modern estimates, the biosphere includes 8.71 ± 1.3 million prokaryotic and eukaryotic species, of which less than 1% have been studied, and only 1.4 million organisms have names (Yablokov et al. 2015).

The body of a living creature contains more than 60 elements of the periodic table, selectively combined in the cells of plants, animals, and humans. A living organism's cell is a living chemical machine. Every second, the human body expels 10 million dead cells. Per year, the total weight of dead cells is equal to the weight of a person. The dead cells are replaced by the newly formed cells at the same level of intensity (Yablokov et al. 2015).

Therefore, it follows that the biosphere is a much more complex system than humans. Therefore, human civilization's control over the biosphere is not realistic and can lead to completely unanticipated outcomes.

The most important fundamental problem of the modern biospheric research is the assessment of the biosphere's ecological capacity that determines the limits of civilization development.

Numerous authors of growth theory, from T. Malthus to its modern followers, consider pollution of the biosphere and the depletion of its resources to be the main limiting growth factors. However, they do not account for the biospheric global eco-logical capacity parameters. And the notion of ecological capacity is often interpreted as the ability of the biosphere to support the existence of the maximum number of people living in a certain territory by using natural resources.

There is another definition of ecological capacity of the biosphere, which refers to the limit area where natural ecosystems (landscapes) can be replaced by artificial (technogenic and agrarian) without threat of global environmental disturbances and an ecological crisis.

An important quantitative indicator of the ecological capacity of the biosphere is the consumption of the net biological production supplied by the biosphere to human population (after using part of resources for its own maintenance). It is estimated that human population can consume about 1% of the net primary production without causing the destruction of the biosphere. In reality, as calculations show (Gorshkov 1980), humans now consume up to 10%; they also destroy about 30% of the net primary production. Thus, the consumption of the net primary production by world population exceeds the sustainable levels by more than an order of magnitude.

There are other theories that interpret differently the role of the biosphere and humans in life on the planet. Some theories represent irrational ideologies that distort the true role of the biosphere. One of such ideologies is F. Bacon's theory of modernism. Supporters of this theory argue that people themselves establish the laws in the biosphere, down to its reorganization. Such a strategy is unacceptable. Humans cannot be held higher than the biosphere that gave rise to them, providing them with nutrition and creating the optimal environment for their existence. The biosphere is incommensurably more complex than human population. The human species is one of the many millions of species of living organisms that exist in the biosphere and, by size, it is an infinitesimal part of the living component of the planet.

Humans as a biological species are "created" by nature with the sole purpose of assimilating the free energy of the biosphere in certain areas of the electromagnetic spectrum. Other organisms assimilate energy in other parts of the spectrum. The free energy of the biosphere, assimilated by the human body, represents the main source of nutrition for all human organs and systems and ensures the vital functions. As soon as people cease to fulfill the main function—to assimilate, i.e., to absorb and bind the free energy of Earth's biosphere, they automatically lose the source of vital energy and transition to the regressive stage of development—aging, which inevitably leads to death (Shchukin 2010).

The thorough study of the biosphere is currently among the main challenges for the scientific community. We urgently need to investigate more rigorously the modern state of the biosphere, its laws, ecological capacity, and extreme (threshold) parameters of technogenesis.

3   How Many of Us Will There Be? Earth's Population

Despite a relative young age of humans as a separate species of biota, the idea of the natural limit of civilization growth is now becoming increasingly widespread. An issue that deserves attention from this point of view is a discussion on three modern global problems capable of disrupting the harmony between human existence and the environment. Such problems include: (a) rapid and uncontrolled growth of human population on Earth and a shortage of natural resources (water, food, and energy) necessary for its existence, (b) development of technogenesis, and (c) global climate change and intensification of catastrophic natural processes.

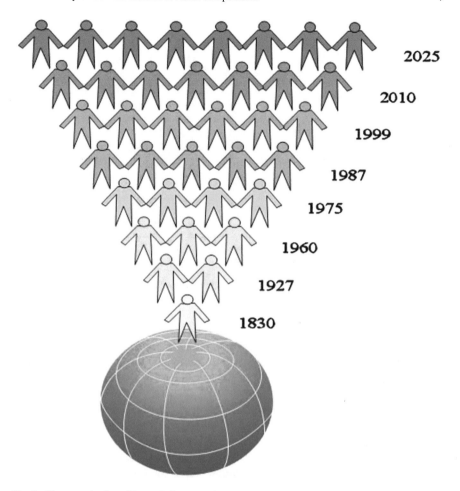

2025

2010

1999

1987

1975

1960

1927

1830

Fig. 1 The growth of world population over the past two centuries

Uncontrolled growth of world population is one of the most important global problems. Over the past 180 years, world population has increased sevenfold and reached seven billion people (Fig. 1). The number of inhabitants of the planet over this period of time has been increasing by one billion people every 10–12 years.

The rapid growth of human population combined with the unbalanced development of the economies of different countries can lead to the most unanticipated outcomes. The flow of migrants to Europe in the spring of 2016 reflects a possible global situation where the lack of employment and livelihoods in the countries with rapid population growth forces people to migrate in search of better living conditions.

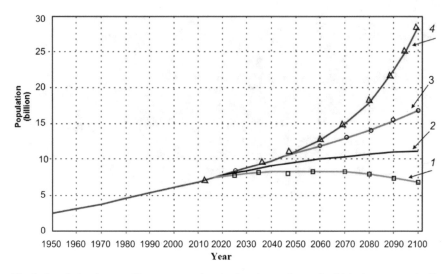

Fig. 2 Population forecast. Forecast scenarios: 1—low, 2—moderate, 3—high, 4—constant fertility (World 2013)

We live in the world where the potential for a catastrophe becomes a reality. It is important to understand the factors that determine population growth, the laws of this growth, and the balance between the natural resources of Earth and the increasing number of its inhabitants.

Demographic forecasting is a very difficult problem, especially when it comes to the planet as a whole. One of the most significant studies in this area is a forecast, based on mathematical modeling, by a Russian scientist Kapitsa (1998). Phenomenological description of world population dynamics has shown that the process of population growth on Earth is uneven: demographic explosion transitions to demographic stabilization, which is observed in developed countries; in the future, this situation can be expected to dominate in developing countries as well.

According to the latest moderate forecast scenario by the United Nations (UN) (World 2013), the population of Earth by the end of the 21st century will asymptotically approach 11 billion (Fig. 2, line 2). This UN forecast scenario is similar to the data of the International Institute for Applied Systems Analysis (IIASA), which project the population of Earth by 2150 at 11.6 billion people. Various scenarios (based on different projections and variants, including fertility rates) suggest a slight decrease in the population on Earth, beginning in 2050 (Fig. 2, line 1), or the continued growth of the population, which by the end of the 21st century can reach 17–28.5 billion people (Fig. 2, lines 3, 4).

Given the continuing general population growth on our planet, it is logical to pose the question of how this growth relates to Earth's natural resources potential necessary for sustainable development of human population of this size. According to some experts, Earth has sufficient natural resources to support 20–25 billion people.

In this case, according to S. P. Kapitsa, the expected limit of the population growth is associated not with external factors, but with internal causes, first of all, restrictions on the rate of population growth.

According to other data, sustainable development of the biosphere is possible if Earth's population does not exceed one billion people. This figure is written in the Declaration "Agenda for the 21st Century," (Agenda 21) (Agenda 1992) adopted at the United Nations Conference on Environment and Development (UNCED) (Rio de Janeiro, 1992). The Declaration states that the planet's "consumer basket" can support approximately one billion people. The concept of the "golden billion" is striking in its unscrupulousness and inhumanity—for its realization it is necessary that about 10 billion people on Earth die of starvation, which is more horrific than the nazi concentration camps and even nuclear war.

Such a large discrepancy in the population estimates based on the consideration of the natural resources of the planet is due to the fact that mathematical models do not adequately account for the influence of technogenesis and specifically of environmental pollution on the frequency of mutations that decrease the viability of living organisms. Accumulation of negative biological changes in living organisms, as a result of changes in genetic control systems and direct influence of chemical and physical factors of environmental pollution, significantly affects human life expectancy and, obviously, the future number of people on Earth.

4 Technogenesis and Its Influence on the State of the Biosphere

4.1 Technogenesis Is the Most Important Factor in the Destruction of the Natural Environment

The present generation of people lives in conditions of technocracy's binging and ruthless consumption of natural resources, without consideration of the consequences.

The industrial and economic activity of humans, termed "technogenesis" by Academician A. E. Fersman, is one of the most important factors affecting the biosphere. The destruction of the biosphere was already taking place in the era of the primitive man—our ancestor Cro-Magnon. But this destruction did not pose a great danger, since it did not go beyond the ecological capacity of the biosphere and did not violate the biosphere's ability to self-repair.

Throughout the 20th century, the world industrial production increased 50-fold. The total consumption increased 16-fold and, by the end of the last century, approached US$30 trillion. In developed countries, the consumption of natural resources by their residents increased hundredfold.

At present, high levels of air emissions from industrial, transport, energy, and social infrastructures result in the pollution of the atmosphere by solid dust particles

(a)   **(b)**

Fig. 3 Emissions of pollutants to the atmosphere: **a** technogenic emissions (photo by V. I. Osipov), **b** emissions from volcanic eruptions

and greenhouse gases. In addition to technogenic emissions, the concentration of CO_2 in the atmosphere is growing due to a reduction in the green cover of the planet and in the absorption of CO_2 by vegetation.

Every year, Earth's atmosphere receives more than 40 billion tons of emissions from stationary and mobile sources, i.e., more than six tons per person per year. This is three to four orders of magnitude greater than the natural emissions from volcanic eruptions (Fig. 3). Especially intensive is the pollution of urban air.

The total world CO_2 emissions from all sources over a little more than the last 200 years (1800–2015) increased from 280 to 400 parts per million parts of air and can reach 800 parts under the most unfavorable scenario by the end of the 21st century; the concentration of CH_4 and N_2O may increase from 0.8 to 1.65 and from 285 to 310 parts per million, respectively.

Chemical anthropogenic pollution of the atmosphere is particularly dangerous because atmosphere receives alien and common chemical compounds in unusually large quantities and forms. According to the World Bank, 5.5 million people died in 2013 worldwide from air pollution related diseases.

As a result of large-scale emissions to the atmosphere (primarily CO_2 and CH_4) and industrial waste, the natural biochemical cycles of carbon, nitrogen, and other elements have changed. The scale of fixation by plants and microorganisms of anthropogenic nitrogen from the atmosphere has increased manyfold. The concentration of phosphorus compounds in Earth's freshwater reservoirs has increased by 75% over the past 50 years (Yablokov et al. 2015).

Processing of mineral resources and accumulation of waste rock in tailing dumps, heaps, and mine takes results in formation of extensive geochemical anomalies of heavy metals, carbon, nitrogen, sulfur, and iodine. The use of mineral fertilizers increases the content of cadmium, arsenic, copper, lead, mercury, and zinc in soils and millions of tons of soluble salts enter soils annually due to irrigation and reclamation of irrigation water.

Besides chemical pollution, there is also anthropogenic physical pollution of the atmosphere, including growing density of electromagnetic and ionizing radiation fields, as well as light and thermal pollution. Sources of such pollution are television, radar-location, high and ultrahigh-frequency currents, and cellular and radio communication.

In addition to atmospheric pollution, technogenesis leads to intense pollution of surface and groundwater and accumulation of industrial waste. According to the latest data from the Federal Center for Subsoil Resources Monitoring (The state report... 2014), the volume of polluted sewage discharge in Russia is about 43 billion m^3 yr^{-1}.

The world ocean receives annually (in million tons): iron compounds—up to 320, phosphorus—22, lead—2, petroleum products—up to 10, and plastic waste—up to 10. In some water areas, the density of plastic waste reaches 14 thousand floating plastic objects per square kilometer (Yablokov et al. 2015).

No less tense situation exists with respect to municipal and industrial waste. Development of new resources and excessive use of the existing have led to the formation of large volume of waste. Annually more than hundred billion tons of mineral substances is extracted from Earth's interior worldwide, i.e., 14 tons per person per year; most (97–98%) of the extracted raw materials is turned into waste. More than five billion tons of waste of various genesis is generated annually in Russia, of which less than half is processed. In total, at least hundred billion tons of waste has accumulated in the country's territory—almost 700 tons per each resident, including children. Such a load of waste on the open surface leads to pollution of the atmosphere, hydrosphere, and lithosphere and it removes almost one million hectares of land from use.

Modern human technogenic activity causes reduction of biodiversity or, according to Charles Darwin, of "the sum of life." About 100–200 species disappear daily. This is three orders of magnitude higher than the natural rate of extinction. According to some data, over the past 40 years, the number of terrestrial animals has decreased by 40% and fresh water species decreased by 75% (Yablokov et al. 2015).

A large number of terrestrial fauna species are under threat of extinction. In the coming decades, elephants, rhinoceroses, polar bears, kangaroos, turtles, tigers, and some kinds of monkeys could be extinct in wild. Already, they are among the unique inhabitants of nature, and in the future, they can be seen only in zoos. The main cause of anthropogenic extinction of species is disturbance of the natural environment of their habitat.

The increased impact of technogenesis has led to the situation when in a number of Earth's regions, the biosphere has exceeded the ecological capacity threshold and is incapable of self-regulation and self-recovery. According to the UN, about 30% of the land surface has already been subjected to environmental degradation due to human activities. In Russia, 15% of the territory is degraded; these are the areas where most of the country's population, production capacities, and most productive agricultural lands are concentrated; the conditions in 20% of the territory are characterized as satisfactory, i.e., changes in the natural environment are observed, but these changes do not exceed the permissible limits; and 65% of the territory of Russia is not much affected by economic activities and retains its biological productivity and biodiversity.

4.2   The Code of Human Technogenic Activity

Thus, technogenesis has triggered irreversible changes in the biosphere and thereby created nearly imminent threat to the existence of human civilization. If the impacts on the biosphere continue at the current level, 50–60% of all species of living beings inhabiting Earth prior to the industrial revolution could disappear by the end of the 21st century.

The potential for biotic regulation over much of the globe is probably still sufficient to compensate for anthropogenic biospheric disturbance and ensure the preservation of the acceptable environment in the designated "corridor" of biotic capacity. To maintain this situation, it is necessary to change fundamentally human economic activities and develop a certain code of technogenesis to regulate the relationships between Humans and Nature. It is imperative to design the necessary steps at the global level to halt further destruction of the biosphere. These steps should be defined under the slogan "*development without degradation of the biosphere.*"

The development of scientific, technical, regulatory, legal, and administrative requirements for economic activities should be considered as the primary step for the implementation of this task. The new code should be based on a principally different technological, moral, and legal fundamental platform that universally incorporates the requirement to prevent further degradation of the natural environment and to use technologies for deep processing and recycling of waste, resulting from the consumption of life-sustaining resources.

Scientifically-based philosophy of biotic regulation of changes in the biosphere should represent the central section of this code. It is necessary to build a quantitative model of the biospheric ecological capacity, which would determine the permissible limits of civilization development and identify the threshold limits of the anthropogenic impact on the biosphere.

The code should specify the standards and prevent exceedance of the threshold impact levels on the environment under various types of economic activities. It may be achieved by implementing various requirements, namely:

- uniform placement of social, industrial, transport, and other facilities to prevent exceedance of the ecological capacity of the individual over-impacted territories;
- elimination of air-pollution emissions not subjected to cleaning;
- prevention of natural water resources use at levels exceeding the capacity of their regeneration;

- prevention of discharge of polluted water into waterbodies without their purification, as well as of forest clearings in the areas with the critical state of the biosphere;
- prevention of excessive use of mineral fertilizers in agricultural production, as well as of use of technologies leading to desertification, salinization, and degradation of soil cover;
- moving away from redistribution of water resources, which violates the natural balance of waters in the regions.

5 "Population Load." The Human Adaptation Corridor

5.1 Technogenesis and Human "Population Load"

Technogenesis not only causes the degradation of the natural environment but also affects humans directly. Chemical and physical factors of environmental pollution increase the frequency of negative mutations that reduce the viability of organisms. The accumulation of negative mutations in living organisms has been termed "genetic load."

The definition of "genetic load" can be considered in a broader sense if both indirect (change of genetic control systems) and direct (toxic) influence of physical and chemical factors of the environment are included. In this case, the notion of "genetic load" can be substituted with the concept of "population load" (Yablokov et al. 2015).

Until recently, human "population load" has not been deeply analyzed, since the average life expectancy of a person increased along with the overall growth of world population. The growth of both parameters was associated with the social factor—the improvement of medical care and the quality of nutrition. Yet, the influence of chemical and physical pollution of the environment on the biological state of the human body was not immediately obvious due to the internal reserves of the human body.

But now, a time is coming when the reserves of the human body enduring anthropogenic influences are exhausted, and human "population load" grows intensely. This can be traced with some parameters of human reproductive capacity. For example, there has been a decline in the number of sperm in young people in recent decades: this decline is 1–2.5% per year in the United States and Europe, respectively. Generalized data show that the number of spermatozoa in men (million/ml) was 80–120 in the middle of the 20th century, while at the beginning of the 21st century, it decreased to 50–70, and it is forecasted to drop to 20–50 by the middle of the 21st century (Nikitin 2005).

The overall increase in the instances of human cancer and mental disorders can serve as the indirect proof of the "population load" growth. Another evidence is the statistics on the poorer health of people living in more contaminated areas compared to less polluted regions.

In addition to technogenesis, the growth of human genetic deficiency is promoted by alcoholism, drug addiction, and overpopulation. Deterioration of the gene pool accompanies the loss of human hereditary and reproductive functions. It is important to note that these evolutionary degradation processes are irreversible, even if we assume that humans could restore the life- supporting functions of the biosphere.

5.2   Adaptation Corridor

The entire history of human civilization represents the continuing adaptation to environmental conditions. The objective laws of the biosphere imply that humans have their own "corridor" where they can adapt and develop. Humans exist only within the tight framework of environmental conditions—within a certain range of temperature, pressure, and chemical composition of air, water, and other components of the environment. Any impact on the environment as a result of the development of natural processes or human activities can lead to the increase in "population load," the disturbance of physical conditions of human existence, and the shrinking of the adaptation corridor.

The average lifespan of a biological species on Earth, according to paleontological data, is seven million years. The preservation of a species is the preservation of its genome, the stability of which depends on the environmental, social, hereditary, and other factors. It was believed that any genotype carried by a species, able to adapt to the environment of the biosphere and produce new offsprings, must survive and ultimately determine the formation of a new species.

In fact, there is no known continuous adaptation leading to the formation of new species in the biosphere. This may be explained by the fact that along with adaptation, living organisms experience various mutations of the genome, which may act as destructive mechanisms that reduce competitiveness and lead to the genome disintegration.

Thus, the adaptation theory turned out to be untenable in relation to humans whose adaptation is substituted with mutation. It is believed that humans have now gone one-third of the way toward the death of the genome (Gorshkov and Makarieva 1997).

The disintegration of the genome is associated with a number of natural and social factors. Unlike most biological species, *Homo sapiens* exhibit a strong social inequality between individuals of the same species. This violates the equal competitiveness between people. Billionaires and millionaires have incomes thousand times exceeding the average incomes of most people, which does not match their real physical or intellectual contribution to the creation of Earth's material values. Deformations in the social structure contribute to the degradation of the human genome. This is manifested in the growth of the number of genetic diseases and is evidenced by the medical statistics of many countries (Gorshkov and Makarieva 1997).

Even more important to the deterioration of the genome is the degradation of the environment and the growth of human "population load." By destroying the external environment and the biosphere, humans automatically narrow the corridor of their existence. Further destruction of the environment will lead to a situation when humans, as a biological species, begin to degrade. This statement is also supported by the forecast of Earth's population for the next one and a half to two centuries, considered above. Ultimately, the biosphere will "get rid" of humans and, according to the law of irreversibility of biological species, humans will never appear on Earth again.

Therefore, in the near future humans have no other way of survival than preserving the environment and the adaptation corridor using the principles of rational nature management. In the distant future, scientific and technological progress and the transition to a new stage in the evolution of the biosphere—the era of intelligence, can establish conditions for a new form of human existence within another adaptation corridor, and most importantly, through a fundamentally different way of obtaining food and other life-sustaining resources.

6 What Happens to Climate?

6.1 Change in Natural Climate Systems in the History of Earth

According to paleoclimatic reconstruction, Earth went through a number of natural climate change megacycles over the course of its history, which were caused by the periodic alternation of prolonged periods of relatively low and elevated temperatures. Each megacycle included shorter climatic cycles. Thus, we can talk about a different periodicity of temperature change on Earth, ranging from several hundred years to hundred million years. This change can be understood only through examination of temperature trends, established by averaging temperatures of shorter cycles. As can be seen from Fig. 4, over 500 million years, Earth had four temperature megacycles with temperature minimums in the Ordovician–Silurian, Carboniferous–Permian, Jurassic–Cretaceous, and Paleogene–Neogene–Quaternary (Kotlyakov and Lorius 2000).

In the Quaternary (the duration of about 1.8 million years) with generally relatively low temperatures, there were several epochs of cooling (glaciation) with interglacial periods of warming. Within the Russian Plain, at least three glaciations with negative temperatures occurred, accompanied by alternating warming epochs with higher temperatures.

The last glaciation took place about 20 thousand years ago. In the subsequent several millennia, warming was observed with a stable trend of a slow rise in temperatures (interglacial stage). In the second half of the Holocene, this trend changed: after the climatic optimum (6000 years ago), slow cooling began (Vakulenko et al. 2006) (Fig. 5).

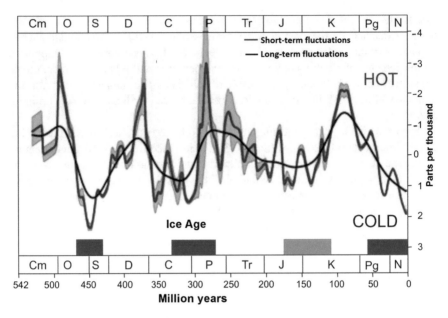

Fig. 4 Temperature change on Earth over 500 million years

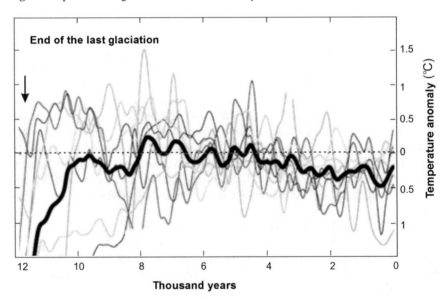

Fig. 5 Change in the average global temperature in the Holocene (last 12 thousand years)

Of special interest is change in global temperature in the last 150 years (Gulyov et al. 2008) (Fig. 6). Meteorological observations show that beginning in the middle of the 19th century, another warming period started and now it has been a worldwide

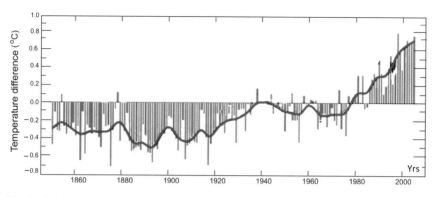

Fig. 6 Global average annual temperature from 1956 to 2002

trend. By 2013, the average global temperature increase reached 1 °C compared with the pre-industrial period, which has led to a marked change in the climatic conditions on Earth. Modern warming is 10 times faster than warming during the transition from the last glaciation to the interglacial.

Detailed quantitative paleoclimatic reconstruction of the last interglacial cycle of the Holocene allowed estimating not only the amplitudes, but also the rates of temperature change (Fig. 7). The rate of change in the mean annual temperature in the second half of the Holocene was 0.002 °C per decade. There was a general cooling trend in the Holocene with climatic fluctuations lasting up to several centuries. The rate of change in the average annual air temperature under such fluctuations was about 0.02 °C per decade. The magnitude of change in the mean annual temperature increased with the decrease in the period of observation. Thus, in 1860–2005 (145 years), it averaged 0.045 °C, while in 1980–2005 (25 years), it averaged 0.177 °C per decade.

An extremely important question arises about whether the current warming is associated with natural climatic variations lasting up for several centuries, which have happened before, or the anthropogenic impact on the climate system is the main factor of warming. In favor of the latter, on the one hand, is a broken correlation over the last decades between temperature and variations in solar radiation, and, on the other hand, the existence of a correlation between temperature and the content of greenhouse gases in the atmosphere.

6.2 Causes of Climate Change

The most important climate-forming factor in the formation of short-term climatic fluctuations is solar activity. Temperature fluctuations correlate with variations in solar radiation on a centennial scale, with the exception of the last decades, when this correlation appears to be broken. A reliable response to solar activity has been

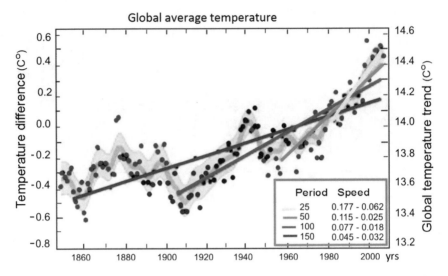

Fig. 7 Change in the rate of global air temperature increase as a function of the observation period

identified in the main climatic characteristics: surface air temperature, ocean surface temperature, and precipitation. The climatic response to the effects of solar and geomagnetic activity is characterized by considerable spatio-temporal heterogeneity and is region-specific (Kotlyakov 2012b).

Comprehensive assessments of the effect of solar activity on temperature change consider the spatio-temporal characteristics of energy exchange between the atmosphere, ocean, and land, as well as atmospheric and oceanic circulation. Regional climate change on Earth is non-uniform.

The world ocean with its "conveyor belt" of currents affects greatly the climate system of Earth's northern hemisphere and the state of the Arctic. The warm surface waters of the Pacific move from the tropics to the Norwegian Sea, warming the Gulf Stream, lose heat, descend to greater depths, and return cooled to the Pacific Ocean (Fig. 8).

Atmospheric circulation has also a significant effect on the spatio-temporal character of climate change. Atmospheric oscillations in the Arctic region cause the formation of polar vortices, the disruption of the jet flow of air masses, and the migration of heavy cooled air deep into the continents (Fig. 9). Atmospheric oscillations are the direct cause of sharp climate fluctuations in the middle and even southern latitudes. The influence of the circulation mechanisms on temperature is nonlinear and can vary with time.

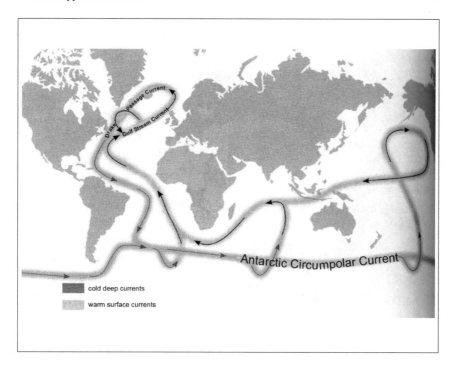

Fig. 8 Global ocean currents and heat and mass transfer in the world ocean

6.3 The Role of Technogenesis in the Global Temperature Change on Earth

The new global climate warming is associated, according to most experts, with anthropogenic factors. Emissions of greenhouse gases (CO_2, CH_4, and NO_2) to the atmosphere contribute to trapping of solar energy in the surface layers of the atmosphere and increase their temperature. Analysis of ice gas bubbles which preserved the composition of the ancient atmosphere of Antarctica shows that the current concentration of greenhouse gases in Earth's atmosphere is much higher than ever in the last 10 thousand years (Kotlyakov 2012a). This gives grounds to believe that the current increase in temperature on Earth is the result of pollution of the atmosphere due to technogenic emissions.

Temperature change on Earth's surface is uneven. The maximum growth of surface temperature is observed in the middle and sub-polar latitudes of the Northern Hemisphere over the continents, where the rate of temperature change can reach 1.7–2.0 °C per century.

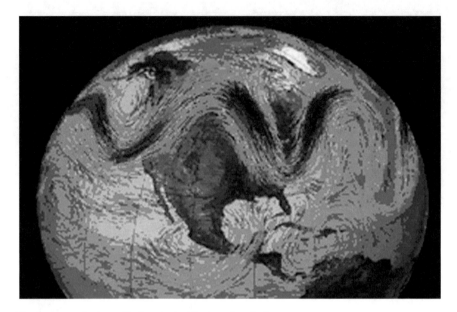

Fig. 9 Atmospheric oscillations in the Northern Hemisphere

Table 1 Increments of the mean annual air temperatures in the north of Russia in 1960–1995

Region	Temperature increments, mean		
	Summer	Winter	Mean annual for 1960–1995
European North	0.4	1.0	0.7
Western Siberia	0.9	1.4	1.2
Yakutia	0.3	2.2	1.4
Northeast	0.1	0.07	0.005

In the territory of Russia, the greatest trend of temperature change was observed in the north of Western Siberia (localities Tazovsky, Salekhard, and Nadym) and in Yakutia (Table 1), while in the extreme northeastern regions, the mean annual air temperature remained practically unchanged.

There are a number of publications both in our country and abroad that describe modeling results of climate change in the last 80–100 years and of its consequences (Mokhov and Yeliseyev 2012). The most authoritative international body dealing with the problems of forecasting global climate change on Earth is the Intergovernmental Panel on Climate Change (IPCC), which issued five evaluation reports (IPCC... 2014). In Russia, the work on climate change is summarized by Roshydromet; this organization has issued two analytical reports (Second Assessment Report... Roshydromet 2014). Below, we provide some actual data contained in the reports of the IPCC and Roshydromet.

Fig. 10 Results of modeling of change in terrestrial temperature on Earth: 1—considering only natural effects (with a constant concentration of greenhouse gases equal to the pre-industrial level), 2—considering change in the observed (through monitoring) concentration of greenhouse gases

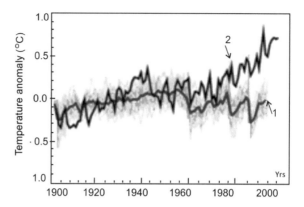

The IPCC conducted a climate change simulation to determine the causes by applying a physico-mathematical model of general circulation of the atmosphere and the ocean. The simulation was carried out for two states of the atmosphere: (1) for a constant concentration of greenhouse gases corresponding to the pre-industrial level and (2) for specific concentrations of these gases according to monitoring data.

Figure 10 demonstrates the simulation results showing change in the mean surface temperature of Earth at a constant (pre-industrial) gas concentration and temperature change with the observed increase in the concentration of greenhouse gases in the atmosphere (Gulyov et al. 2008).

Based on the data obtained, the IPCC stated that it is possible, with a high degree of probability (90%), that the temperature change observed over the past 50 years is caused not only by natural factors, but also by the increase in the concentration of greenhouse gases. Only taking this factor into account, one can confidently reproduce trends in the mean global temperature in models.

An important conclusion of the IPCC Fourth Assessment Report (IPCC... 2007) was that the anthropogenic impact on climate is manifested in all inhabited continents not only in change in air temperature, but also in the characteristics of atmospheric circulation and in the increase in the frequency of catastrophic natural phenomena.

It also follows from the IPCC reports that in the next two decades, regardless of greenhouse gas emissions scenario, global warming will continue at a rate of about 0.2 °C per decade. Even if greenhouse gas emissions do not grow, the temperature increment of 0.1 °C per decade should be expected in the next 20 years. This suggests that with greenhouse gas emissions at the present level and especially with their increase, there is a high degree of probability that warming on Earth, change in the global climate system, and the development of catastrophic natural phenomena accompanying it will continue further. In accordance with the climate model of A. M. Obukhov Institute of Atmospheric Physics (Russian Academy of Sciences [RAS]), global warming in the 21st century will be 1.1–2.9 °C on average, depending on scenario (Mokhov and Yeliseyev 2012) (Fig. 11). Temperature will rise particularly rapidly in the Arctic and on the continents. In this regard, we should expect increase in the level of the world ocean and the degradation of polar ice and permafrost.

Fig. 11 Change in the mean global air temperature in the 21st century under different scenarios of the anthropogenic impact on climate: mild (1), moderate (2), and harsh (3)

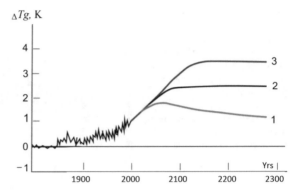

Obviously, the consequences of climate change will be mixed and will be both positive and negative. The positive consequences include improved conditions for the development of the Arctic shelf, shorter heating season, energy conservation, shift to the north of the zone of risky farming, etc. There is a large number of negative outcomes, primarily, increase in the number of dangerous natural processes. In this regard, the problem of adaptation of all economic and social activities to the expected climate change becomes particularly important. It is crucial to remember the fundamental principle recorded in the Climate Doctrine of the Russian Federation, *"Losses come to us easily but the use of benefits requires us to make an effort."*

Delays in adaptation measures will result in higher incidents of emergencies associated with great economic costs and increased risk of human loss. Therefore, a proactive early adaptation program is needed, whose main objective is to increase sustainability of society and its resilience to changes that are taking place in the environment.

Goethe wrote, *"Nature understands no jesting; she is always true, always serious, always severe; she is always right, and the errors and faults are always those of man."*

7   Life and Natural Disasters Risk

7.1   Intensification of Natural Disasters

Catastrophic natural phenomena are considered among the most important destabilizing factors that cause great damage to the economy and the environment.

In the second half of the 20th century, the number of natural disasters increased dramatically. According to the data of the Centre for Research on the Epidemiology of Disasters (CRED) (http://www.emdat.be) for (1980–2015), the annual number of natural disasters over 35 years increased almost threefold (Fig. 12). According to

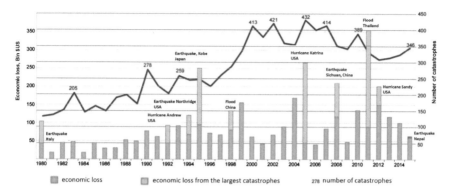

Fig. 12 Natural disasters and the magnitude of damage on Earth in 1980–2015

the UN, over the past 20 years, there were 7056 natural disasters in the world, in which 1.35 million people were killed and 4.2 billion people were injured (Delovaya Gazeta 2016).

The spectrum of natural disasters is exceptionally large—from seismic phenomena to widely occurring floods, droughts, slope processes (landslides, mudflows, avalanches), storms, hurricanes, karst collapses, sinkholes, erosion, abrasion, etc.

The rapid growth in the number of dangerous natural processes is due mainly to three global processes: internal geodynamics of Earth, development of technogenesis, and climate change.

7.2 Internal Geodynamics of Earth (Seismic Phenomena)

V. I. Vernadsky wrote, "*Earth cannot be regarded only as the area of matter, it is the region of energy.*"

Internal geodynamics of Earth is associated with processes developing in its crust and upper mantle, which cause changes in the stress state of the outer shell of Earth and the development of deformation processes. The latter is manifested in tectonics and seismicity of Earth. The greatest danger is associated with seismic processes that can develop rapidly and unexpectedly, be accompanied by release of a large amount of energy, and lead, as a rule, to major disasters on Earth's surface.

There have been exceptional, in terms of energy and social consequences, earthquakes in the history of Earth. Damages from the largest earthquakes are comparable to the most destructive wars. Analysis of the available data indicates that in the second half of the 20th century, the number of large earthquakes on Earth increased significantly (http://www.emdat.be/) (Fig. 13).

The reasons for such growth of seismic phenomena are not yet completely understood. Perhaps they are associated with a new phase of Earth's endogenic processes intensification. Technogenesis may affect the development of so-called induced

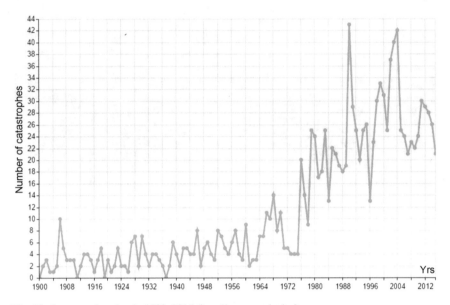

Fig. 13 Large earthquakes in 1900–2015 (http://www.emdat.be/)

Table 2 The most destructive earthquakes on earth in 1999–2015

Date and country	Magnitude (Mw)	Depth of focus (km)	Number of deaths, people	Number of injured, people
August 17, 1999, Turkey	7.58	15	17,127	1,358,953
May 21, 2003, Algeria	6.81	12	2,266	210,261
December, 26, 2003, Iran	6.59	15	26,796	267,628
December, 26 2004, Indonesia	9	30	165,708	532,898
May 12, 2008, China	10	7.91	87,476	45,610,000
April 6, 2009, Italy	10	6.32	295	56,000
April 25, 2015, Nepal	8.2	7.8	8831	5,639,722

earthquakes. Some data suggest that construction of dams with a height of more than 140 m, in 21% of cases, leads to induced seismic phenomena with magnitudes in the range 5–6. The impact of underground nuclear explosions on induced seismicity has also been recognized.

The most destructive seismic events of the recent years (1999–2015) and the social and economic consequences caused by them are presented in Table 2.

Fig. 14 The earthquake and tsunami in Japan (March 2011), which led to the accident at the Fukushima nuclear power plant

Seismic phenomena are associated with a number of major disasters that represent a global threat. Among such accidents, for example, is the destruction of the Fukushima nuclear power plant in Japan. The station seismic design was appropriate for the maximum magnitude 7 earthquake. The accident occurred on March 11, 2011, after a magnitude 9 earthquake. The earthquake caused a 14-m high tsunami (Fig. 14), which greatly exceeded the permissible height of 6 m in the project design. The tsunami wave flooded four out of the six station's blocks, disabling the reactor cooling system. This led to a series of hydrogen explosions, fires, melting of the core, release of radioactive substances into the atmosphere, and the death of more than 20 thousand people.

No less terrible tragedy happened in December 2004. A magnitude 9 earthquake that occurred in the Indian Ocean caused huge tsunami waves that reached the coasts of Indonesia and a number of other coastal states. More than 165 thousand people of these countries died. The largest seismic event in the 21st century was the Wenchuan earthquake in China on May 12, 2008 (magnitude 10) (Figs. 15, 16).

Fig. 15 Earthquake damage in the city of Beichuan; 80% of buildings were destroyed

7.3 Dangerous Technogenic Processes

Intensification of natural disasters in recent years is caused not only by natural factors, but also by the growth of dangerous phenomena triggered by human activities. Such phenomena include: karst collapse, flooding, and slope processes (landslides, mudflows, etc.).

Dangerous technogenic processes can be of a significant scale and pose a serious threat to people and infrastructure. Figure 17 shows the Berezniki industrial region (Russia) karst collapse that occurred in July of 2008. The development of potassium salt deposits in this region by large-volume underground workings (up to 80 million m^3) had weakened the massifs of salt and overburden layers and reduced their density due to the hydrodynamic and geochemical impact of groundwater, which led to deformations, destruction, and flooding of the excavation sites. This was the reason for the development of the karst process and the extensive collapse. The collapse was 300 × 400 m wide, 80 m deep, and with a volume of 8.6 million m^3.

The risk of further development of catastrophic geological phenomena in the residential areas of Berezniki, located above the mine workings, is being assessed. One of the response options considered is the relocation of Berezniki to a safer territory.

Fig. 16 The city of Beichuan after the earthquake

7.4 The Effects of Global Climate Change

Among the factors that cause intensification of natural processes, an important role belongs to global warming. The observed temperature rise is accompanied by dangerous natural processes such as drought, acid rain, melting of Arctic ice, degradation of permafrost, and intensification of geological and especially of hydrometeorological catastrophic phenomena. In 1991–2010, their number increased 3.5-fold. In 2015 alone, more than 30 large-scale droughts and catastrophic floods were recorded worldwide, with 98.6 million people affected.

A major flood occurred in 1998 on the Yangtze River, which became the most devastating in the history of China—223 million people were impacted and economic damage amounted to US$36 billion.

Among the extreme hydrological events that occurred in recent years in Russia are the spring flood on the Lena River in Yakutia in 2001, the flood in the Krasnodar Territory in June 2012, and the flood in the Far East in 2013 (Fig. 18). The latter is considered particularly devastating—235 settlements were flooded in 37 districts of the Far East; 678 thousand ha of agricultural land, 430 km of automobile roads, and 71 bridges were destroyed or flooded; 135 thousand people were affected. The total damage amounted to 258 billion rubles (Porfiryev 2015) (approx., US$4 billion).

Fig. 17   The Berezniki industrial region karst collapse

Abrupt climate change events have caused intense growth of catastrophic meteoro-logical phenomena, such as extreme heat and forest fires in Europe and the European part of Russia in 2010 and an unusually cold winter of 2012.

The role of technogenesis in climate warming is particularly evident in the Arctic region and the degradation of permafrost. A. M. Obukhov Institute of Atmospheric Physics (RAS) carried out a model assessment of the degradation of permafrost under global climate change; the assessment considered the anthropogenic impact. Accord-ing to the calculations, the total area of near-surface permafrost soils of the Northern Hemisphere in the 21st century may decrease from 20 million km^2 to 5.3–12.8 million km^2, depending on anthropogenic impact scenario (Fig. 19) (Mokhov and Yeliseyev 2012).

The reduction of the cryolithozone area is caused by the thawing of near-surface frozen soils in the northern regions of the central part of Russia, Western Siberia, and North America. Under the most severe anthropogenic impact scenario, some parts of Eastern Siberia may be affected by the degradation also. In the persistent permafrost regions, the depth of seasonal thawing may increase. Thus, in Eastern Siberia, it may increase from 2.0–2.5 m (under the moderate anthropogenic impact scenario) to 2.5–3.0 m (under the harsh scenario).

Fig. 18 Floods in the Far East (2013): **a** Magadan Region, **b** Amur Region, **c** Khabarovsk

Fig. 19 Change in permafrost soils area in the 20th–21st centuries under mild (1), moderate (2), and harsh (3) scenarios of the anthropogenic impact

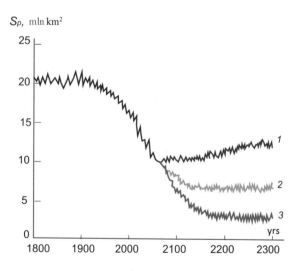

Air pollution of the Arctic is causing intense melting of polar ice. For example, the emissions of the Norilsk Combine (Russia), whose main pollutant—the aluminum plant, was recently closed, were registered at the North Pole.

Fig. 20   An outcrop of sheeted ice on the Yamal Peninsula

The area of pack-ice in the Arctic has shrunk to half of its size over the past half century. This process is uneven and most clearly manifested in the zone of influence of the Gulf Stream warm current and is practically silent in Eastern Alaska and Northern Greenland.

The process of glacier shrinkage in the Arctic is not less intensive. A forecast assessment of the state of the Russian Arctic archipelagoes glaciation over the next decades shows that the deficit in the mass balance of glaciers will increase in the current century. Especially high specific mass losses should be expected on the glaciers of Franz Josef Land and Novaya Zemlya.

The transition of soils from frozen to thawed conditions leads to deformations of Earth's surface and the development of dangerous natural phenomena in vast areas of Western and Eastern Siberia. The most important destabilizing factor in this region is thawing of soils in the area of distribution of ice deposits, polygonal vein ice or ice wedges, and sheeted ice. This process in the continental part of the Arctic enhances thermokarst, thermoerosion, solifluction, seasonal heaving, and the subsidence of Earth's surface.

The most significant changes should be expected in the regions of soils and rock strata with large ice volume. These territories include the Yamal Peninsula, where the largest Bovanenkovo gas condensate field is located; there, ice sheets with an average thickness of 8 m were encountered (the maximum thickness is 28.5 m) (Fig. 20). The area of some ice inclusions reaches $10\,km^2$, and the volume—more than 4 million m^3.

The widely occurring processes associated with permafrost degradation include thermokarst. The mechanism of formation of karst sinkholes (lakes) is usually the

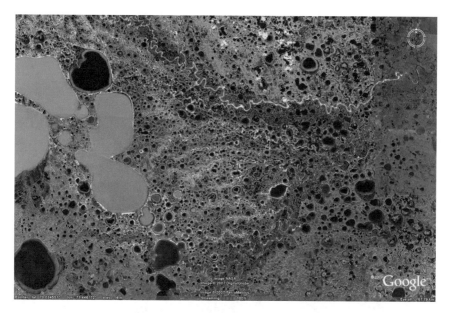

Fig. 21 A region of the Yamal Peninsula affected by thermokarst (space imagery)

result of the collapse of the roof of rocks that overlap cavities (voids) arising in the frozen strata due to local thawing of buried ice or ice-saturated soils. The severity of thermokarst in the northern regions of Western Siberia is extremely high (Fig. 21).

Recently, another thermokarst formation mechanism—pneumatic or explosive, has been discovered in the north of Western Siberia. Thus, in July of 2014, a deep crater that resembled an explosion funnel and filled with water at the bottom, was found 30 km south of the Bovanenkovo oil and gas condensate field in the south-western part of the Yamal Peninsula (Bogoyavlensky 2014) (Fig. 22).

The funnel has a rounded shape with a diameter on the surface of about 60 m and a depth of more than 50 m. It is surrounded by a bank made of earth ejected to a distance of up to 120 m. The general appearance of the funnel indicates that it was formed as a result of a powerful release of gas from a shallow subterranean deposit formed, possibly, as a result of thawing of buried ice (stratified, vein or bulgunyakh nucleus–hydrolaccolithes) and the accumulation of gas. The accumulated gas in the cavity could be syngenetic (of biochemical origin) or catagenetic, migrating from deeper horizons through deep-seated faults. Its formation may be also the result of dissociation (decomposition) of gas hydrates when thermobaric conditions change.

The described gas-explosive funnel is not the only example in the Arctic zone. Other similar formations were found in Yamal and in the mouth of the Yenisei River.

Explosive degassing of the cryosphere also occurs on the Arctic shelf where a submarine permanently frozen layer is present. This is evidenced by deep craters on

Fig. 22 The thermokarst crater (explosive type) on the Yamal Peninsula. A helicopter view

the seabed of the shelf zone. Explosive methane emissions pose a great danger for boreholes, underwater pipelines, and vessels. In the latter case, a vessel can rapidly sink, since the density of water in the methane emission zone is reduced by half, and the vessel cannot remain afloat. There have been four tragic events leading to the total destruction or severe consequences for ships and their crews (Vinogradov et al. 2016).

7.5 Socio-economic Losses

Human-induced and natural disasters on Earth cause huge social and material losses. According to the Centre for Research on the Epidemiology of Disasters (CRED) (http://www.emdat.be/), a significant part of the total annual damage (from 15 to 55%) is associated with seismic phenomena. Floods, waterlogging, droughts, landslides, mudslides, avalanches, and other dangerous phenomena also contribute significantly to the total damages.

Seismic events are associated with the largest number of deaths in world catastrophes (about 57% of the total), followed by floods, storms, and geological hazards. Floods cause the most damage (up to 56% of the total), followed by droughts, earthquakes, and geological hazards.

The total annual socio-economic damage from all kinds of catastrophic phenomena on Earth in 1980–2015 varied from US$120 billion to US$350 billion (Fig. 12), gradually increasing with time. The annual global damage peaked in 2011 at US$350 billion. In subsequent years, the damage was somewhat lower. However, at the Third World Conference on Disaster Risk Reduction, held in Sendai, Japan in March of 2015, where 170 countries participated, UN Secretary-General Ban Ki-moon stated that the annual global damage from natural disasters will increase to US$350 billion in the coming years; by 2030, it could rise to US$360 billion.

Some countries, for example Japan, spend as much as 5% (and even 8%) of their annual budget to address the consequences of natural disasters.

7.6 Assessment of Natural Risks

Natural risk is the probabilistic value of possible damage (social and economic) from hazardous natural phenomena in specific areas.

Natural risk is a function of the probability of natural hazards occurring in the territory under consideration and the vulnerability of objects under risk located in this territory. The objects under risk are considered to be all objects to which the concept of damage may be applied.

Natural hazard [H] is estimated as the probability of occurrence of a dangerous process of a certain intensity in a given territory at a given time. Therefore, natural hazard is estimated as the probability of occurrence of a process, of a certain intensity per unit area per unit time.

Vulnerability [V(I)] is a function linking the loss of properties by objects under risk and the strength (intensity) of a hazardous natural process.

Risk [R] can be estimated from the values of H and V(I) as follows:

$$R = H \times V(I).$$

Thus, risk assessment includes superimposing information layers on natural hazards and vulnerability of certain areas and obtaining integrated information as the product of the two variables. In practice, this is reduced to the procedure of compiling a forecast map of the frequency of development of a dangerous process for the study area and overlaying this map over a map on settlements and objects of the technosphere, with concurrent assessment of their conditions (vulnerability) and urbanization characteristics (Fig. 23). The resultant risk map identifies territories and settlements with different probabilities of individual death and injury or possible material damage within a specified period of time.

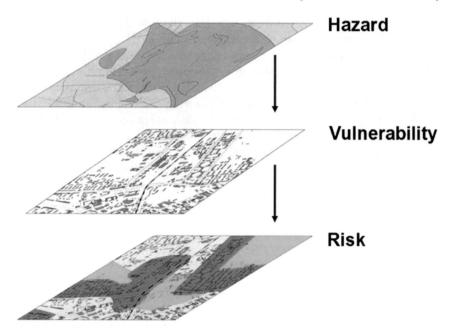

Fig. 23 Information layers for obtaining a risk map

This procedure is used to target a number of tasks and assess the existing risks. The end result is the spatio-temporal probabilistic-deterministic forecast of the consequences of various natural risks, expressed as the estimated probability of an individual or integral damage (death and injury to people), as well as of losses in the social and material sectors (Larionov and Frolova 2005).

Russia has accumulated considerable experience in compiling natural risk maps. There are risk maps of the most probable and destructive natural hazards (e.g., seismic or multi-hazard–seismic, landslide, flood, karst, etc.) for the entire country, individual regions, and municipalities (Osipov et al. 2011).

The most common are maps of individual natural risk reflecting the probability of the individual natural death risk of a person, maps characterizing human death and injury, and integral maps of death and injury to people and material damage.

As an example, Fig. 24 shows a risk map compiled by the staff of Sergeev Institute of Environmental Geoscience (IEG RAS) in cooperation with the Extreme Situations Research Centre. The map reflects the probability of integral natural risk (death and injury) to individuals in Russia from the following hazards: earthquakes, floods, hurricanes, avalanches, landslides, and mudflows.

Risk (fatalities and injuries of varying severity) from six types of natural hazards varies from negligibly small (close to zero) to very high (of over 100×10^{-5} yr^{-1}) values.

Figure 25 presents an example of maps of individual risk for certain types of hazard; it shows seismic risk for the Krasnodar Territory.

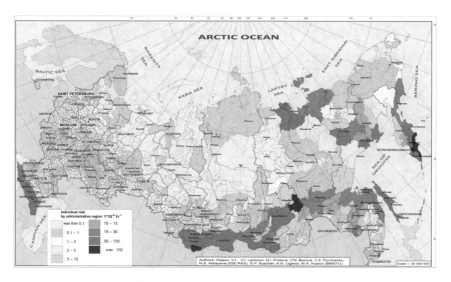

Fig. 24 Map of the integral natural risk (death or injury) to individuals in the territory of Russia based on the probability of occurrence of earthquakes, floods, hurricanes, snow avalanches, landslides, and mudflows

The risk assessment procedure may be applied to compare different areas to find the safest, to identify the optimal options for their settlement and use for facilities of the technosphere, to design natural risk reduction measures, and to optimize the Russian Ministry of Emergency Situations rescue equipment placement in the areas under increased natural risks. For example, in 2007, an individual risk assessment was carried out for the city of Greater Sochi; it was estimated at 11.6×10^{-5} yr^{-1}, which exceeded the standard regulatory value. Subsequently in the course of preparation for the Olympic Games in 2014, dilapidated buildings in the city were demolished and replaced with earthquake-resistant structures. As a result, the level of seismic risk was reduced threefold to 3.66×10^{-5} yr^{-1} (Osipov et al. 2015). If the rate of earthquake-resistant construction envisaged by the city plan is met, by 2032, the average normalized value of seismic risk for the whole city will be reduced ninefold, compared to 2007, and will reach the acceptable level of 1.3×10^{-5} yr^{-1}.

8 Market Economy and Ecology

Currently, the economic development strategy of the most countries in the world is based on the concept of the market economy. Numerous declarations, statements, and memorandums praise the advantages of this type of management over other strategies. The arguments suggest large potential of the market economy, its efficiency, business affability, liberalism, etc.

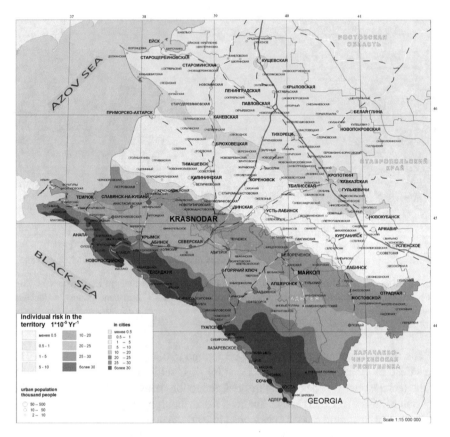

Fig. 25 Map of individual seismic risk for the Krasnodar Territory

   The market is a universal selection mechanism operating in nature and society. It acts as a complex and hierarchically organized system of rejections of old structures and their replacement by new continuously created structures. The market is the area of spontaneous self-organization—it selects the most adaptable and successful entities in specific conditions. At the same time, the market often makes mistakes, because it is incapable of taking into account the prospects and general trends of development. Therefore, the market acting in nature and society is the most complex entanglement of various contradictions.

   Success of the market economy dampens peoples' attention to the long-term prospects of the existence of future generations. Along with national security, their future will be determined by the environmental conditions and environmental security of the country. In this work, we would like to discuss from the position of common sense how the market economy affects environmental security and what the future of human society is if it continues to adhere to the principles of the market economy.

We do not question the works of economists and various economic theories on the market economy and its dominance. We would just like to draw attention to the practically concurrent challenges to environmental security. There is a concern that if society continues to develop along the accepted trajectory, the onset of a global ecological catastrophe may be accelerated, from which no economy, even the market economy, can rescue the situation.

Nowadays, we are facing an increasingly frequent situation when successful public or private companies that bring significant income to the budget of the country and regions are also the largest environmental polluters and thus their activity causes various environmental problems. There is a conflict of interest between these companies and the residents of the regions. More precisely, there is a conflict between business and the interests of society.

The strategic basis of any business is achieving the maximum profit. Environmental protection measures and the transition to environmentally friendly technologies require additional costs and thereby reduce the efficiency of the business. Companies working in the mining sector often switch their operation to new areas as soon as possible and start developing them using traditional methods instead of carrying out a more complete extraction of raw materials in the already developed fields using modern technologies, including deep processing of waste and recycling. This also applies to social problems, because they, like environmental, create additional overhead.

The modern market economy functions as a powerful machine that destroys the biosphere and the spiritual and physiological essence of humans. Undoubtedly, it is a powerful source of innovations that contribute to the prosperity of a small part of world population (the "golden billion"), while the biosphere and the rest of population are under severe deteriorative impacts from business activities in general.

Governments and business often intentionally overlook this problem and elevate the role of the market economy. Supporters of the market economy and sociologists believe that it creates a mechanism for self-organization of society, where social issues and the problems of the biosphere are included in the economy.

Many apologists of market freedom and private property see them as guarantees of democracy, humanism, and human rights. It is hardly possible to agree with this notion.

As it turned out, the market strategy is "blind" and is focused only on the growth of profit, which is why it is far from comprehending the problem of human survival in the conditions of the deteriorating biospheric regulatory capacity.

From what has been said, one can draw the obvious conclusion—the market economy is not compatible with ecology. Such a statement cannot be viewed as the author's firm position. This cannot be regarded as the condemnation of the market economy, because there are many opportunities to reverse the growing antagonism in society. It is necessary to switch to fundamentally different relations between nature and society. Therefore, the author sees the main task of this work in the search for a convergence of the interests of ecology and the market economy.

The solution to this problem requires a policy mutually beneficial for entrepreneurs and residents of the country. The users of natural resources should realize that natural resources are the property of all people and even privatization does not mean the full rights of owners to treat resources without regard to the interests of the country's citizens.

The conclusion drawn above does not mean necessary sharp reduction in the consumption of natural resources to preserve the biosphere. This means that humans must find a new form of economic activity which would allow them to grow the economy and preserve the biosphere and its life-sustaining resources. Humans have now reached a level of technological development that allows effectively balancing environmental security and economic growth. The strategic goal of modern civilization is the inclusion of economic activity in the modern integrated process of the development of society and nature.

9 Protection of Nature and Nature Management. Environmental Security

The interaction of humans with the environment is diverse and versatile. The concept of "nature protection" is often used in discussion of this problem. However, this term does not completely cover the essence of our relationship with nature; it is much more complicated than just protection. Thus, two terms—protection (conservation) of nature and nature management, should be used in addressing the issue of interaction between society and nature.

9.1 Conservation of Nature

The concept of *conservation* or *preservation* of nature has a wide range of meanings; first of all, it means the conservation of biodiversity and its careful treatment. This concept is based on the main idea of maintaining the existing ecosystems and preventing their degradation. An important factor in its implementation is careful treatment of living organisms (including humans) and their environment.

The concept of careful treatment of biota and the environment is of the intellectual nature and is determined by two categories of human intelligence—environmental education and environmental upbringing. The first notion refers to the important but not a defining category. More important is environmental upbringing which has an inner subconscious character, close to spiritual. It should be treated as a genetic, hereditary, concept. Often people who do not have any ecological education, but who advocate goodness and love to humanity, treat nature with due care.

9.2 Nature Management

The problem of interaction between nature and society is not limited to the conservation of biodiversity. From the very beginning, humans used natural resources to obtain necessary means of subsistence and this was the basis for their survival and development. Thus, the structure of nature management on Earth emerged. Currently, the process of nature management is developing in two directions: (a) consumption of life-sustaining resources and (b) use of natural resources for comfortable and safe existence on Earth.

The first direction, i.e., *the consumption of life-sustaining resources*, is the vital need of the use of the biological and mineral resources of the biosphere for nutrition, energy, infrastructure, etc. Humans cannot exist without the consumption of natural resources. In most cases, this consumption is irreversible; the consumed life-sustaining resources are not restored, or their recovery lags behind the consumption, which ultimately leads to their gradual depletion in the biosphere. This phenomenon is observed, for example, in extraction and use of oil, gas, coal, water, and some sources of plant and animal food.

Impending and continuously increasing shortage of life-sustaining resources, caused by technogenesis and population growth, necessitates transition of modern civilization to *rational nature use*. Humans must stringently account for the consumed products, avoiding their unnecessary over-use due to low culture of extraction, processing, and utilization of raw materials, so that resources of nature are sufficient for the present and future generations.

Rational nature management is, first of all, a technological problem. Its solution is based on the application of methods and techniques that optimize processes for the development of natural resources, their integrated and deep processing, secondary use (recycling) of waste, etc. Such technologies include:

- consumption of biospheric resources with the maximum and integrated extraction of useful components and deep processing and utilization of waste as a secondary raw resource for its further economic use;
- application of energy-saving technologies based on highly efficient and complete (with minimum emissions) burning of organic raw materials in combination with the use of solar, water, wind, and other environmentally friendly types of energy; and
- improvement of nuclear energy production with a focus on complete combustion of the radioactive component, further processing and use of the resulting decomposition products, and their transformation into radiation-safe elements.

However, technological approach to rational nature consumption is not a "panacea" for future problems. Rational use of natural resources can prolong the existence of human civilization, but it cannot divert the threat of its collapse as a result of complete exhaustion, in the long term, of Earth's life-sustaining resources.

Therefore, another strategy of rational consumption of natural resources has become an increasingly frequent topic in discussion. It is based on the concept of *regeneration* of life-sustaining resources. Such an approach assumes recycling of generated waste, its processing, and decomposition, so that the resulting chemical elements are returned to the natural biogeochemical cycles.

The return of chemical elements of life-sustaining products to the biospheric biochemical cycles is the most important problem of human survival. This concept fits with ecosystem functioning and assumes that an adequate amount of substances necessary for the synthesis of new life-sustaining products is produced from waste processing. Systematic research in this direction is just beginning. The implementation of the regeneration strategy constitutes the first and foremost priority in human civilization survival and its success will be determined solely by the level of science and technology.

The second direction of the use of natural resources, which is evolving now, can be called *nature management based on the adaptive principle*. Humans exist in the surrounding natural environment and not only consume its resources for survival, but also use them to make their existence comfortable and safe. Such actions can be called adaptive nature management, which means that humans strive to adapt to the conditions of the natural environment, i.e., "inscribe" their existence and activity in natural conditions and processes.

In this context, the concept of "adaptation" in nature management means adjusting social and economic activities of humans to natural conditions to achieve the greatest effect, while maintaining or minimizing the transformation of the natural environment.

The simplest adaptation mechanism is the agricultural activity based on the conscious use of soil fertility and climatic conditions for cultivation. Adaptive nature management is also applied in forestry, water management, construction, etc.

Adaptation is an effective co-evolutionary mechanism of nature management, which is fundamentally different from just consumption of natural resources. At the heart of adaptive nature management are mutual adaptation and the basic principle—humans should not alter natural processes and provoke their negative evolution. This means that human economic activities should not violate the laws of nature.

Adaptive nature management utilizes broadly nature-like technologies. Application of nature-like technologies is facilitated by territorial regionalization maps (seismic, engineering-geological, landscape, etc.) used for socio-economic optimization of placement of cities, settlements, and entities of the technosphere, with provisions for securing nature, ecological capacity of the biosphere, and attractiveness of the natural environment.

Nature-like technologies also include assessment of natural hazards and risks, which supports the implementation of hazard-and-risk reduction measures and creation of management systems for preventing and eliminating emergencies. One example of such technologies is the map on the seismic micro-zoning of the Imereti Lowland on the Black Sea coast compiled for the construction of the Olympic sports complex. It appeared that within a relatively small area of the lowlands, seismicity varies in the range of magnitude 7–9 (with a frequency of 500 years). This map allowed placing sports facilities on sites with the least seismicity thereby increasing safety and optimizing construction costs (Osipov et al. 2015).

9.3 Environmental Security

The strategy of nature protection and nature management is closely connected with environmental security. *Environmental security* assumes protection of the biosphere and hence of humans, society, and states from threats caused by negative changes in the environment under the influence of economic and other activities of people, as well as of natural phenomena.

For the people of many countries, the most important is the issue of military, not environmental security, because there continuously exists the threat of the global war. Environmental security seems more remote and therefore less threatening.

The formation of such a worldview in modern society is largely facilitated by the policies of individual countries and governments, aimed exclusively at the arms race and economic rivalry for the sake of enrichment and world domination.

The current prioritization of the goals for ensuring global security cannot be considered correct, if only because *environmental insecurity* can be more massive and the consequences can be more painful than even nuclear war. It is also necessary to consider the degree of public preparedness to address these problems. Ensuring military security is a purely political problem—it is possible for national leaders to work out a unified decision on the need for disarmament and elimination of military threat; this can be carried out, since there are no technological obstacles to the implementation.

It is much more difficult to guarantee environmental security, because it would be necessary to know the laws of harmonious development of nature and society and then use this knowledge to develop scientific and technological approaches to minimize the global ecological threat. It would involve development of specific approaches for alleviation of environmental tension on Earth, change of technogenesis strategy, and solution of many other problems. Modern society does not yet fully possess such knowledge and technologies, and therefore is not ready to solve these problems. This requires time and concentration of all the efforts on the global scale and overcoming existing unresolved problems. Any impediment in the solution of these problems and conscious unwillingness of the public to understand them makes the ecological threat inevitable and catastrophic.

In a number of countries, including Russia, measures are being taken to develop the national strategy for environmental security. Unfortunately, most of the strategies are anthropo-centric and focus on achieving a unified state policy which is a set of coordinated political, legal, control, supervision, socio-economic, organizational, information, and other measures. However, the fundamental problems of the biosphere and the patterns of interaction between society and nature remain outside the framework of these strategies.

There exists the anti-science and anti-human "golden billion" theory which is a common topic of discussion subject in the media. This theory is based on the idea that the depletion of natural resources and the deterioration of the environmental situation on Earth will inevitably lead to massive migration of people in search of favorable territories for life and the gradual extinction of world population. Naturally, extinction will begin with the poorest segments of the population. Sociologists even calculated that the extinction can reduce world population to one billion. The remaining wealthy population will continue to exist safely, as Earth's resources will be sufficient.

The strategy of the "golden billion," striking with its cynicism, does not hold water because, according to the laws of evolution of the biosphere, disappearing or endangered species on Earth cease to exist completely over time, as they lose their adaptive capacity for further survival in the changed environment. Therefore, this scenario, most likely, is not realistic and if the ecological crisis aggravates further, the human race on Earth will most likely disappear completely.

10   Nature as a Living Soul. Ethical Aspect

10.1   Life with Nature

There is nothing more imaginative, more diverse, and more poetic in the world than nature. Love for nature encourages love for humanity and the surrounding world, makes life interesting and multifaceted, and fills it with meaning. People who live in close contact with nature become dreamers; their existence is enriched by understanding of the essence of life. The scent of nature, its light, and fragrant tenderness fill the soul of everyone who lives the full life. Happy are those who can see and feel, as nature awakens in the spring and withers in the fall, can feel what air, water, grass, forests, and mountains are, and can enjoy the complete mosaic of all this, called the landscape. In the spring, it is touching to watch how nature carefully covers dead grass and dead branches of trees with young green shoots and everything is transforming again—the triumph of life is everywhere. Nature is ever-changing. A coming spring never again repeats the previous spring, and an awakened brook tries to make its own way on Earth.

Nature inspires humans to creativity. It is not by chance that it is the source of poetic muse, it does not leave many poets, writers, and other creative people apathetic.

The moral animation of nature is reflected in the poetry of F. Tyutchev:

Not what you think the Nature is
− an apathetic face or cast.
It is with soul and full of freedom,
it speaks a tender tongue of love.

Metaphorically speaking, nature can be a partner in conversation. There is a well-known phrase, *"The artist knows that Earth is alive, and Nature is divine."*

10.2   A Live Drop of Water

Many elements of nature inspire us; they become inseparable from the concept of life on Earth. Prishvin (1983) wrote a magnificent story "A Forest Water Drip," where, with the inherent spiritual mastery of this writer, the inner secrets stored in a drop of forest moisture are described. The unique beauty of this description moves us deeply and ultimately makes us think that this should be preserved forever and touch future generations. One has only to remember the words, *"In the early spring, in undressed forest, drops of moisture on the branches of trees glow like sparks, whose light illuminates the spring morning."*

The removal of water from biota leads to the degradation and death of the living, and then of everything else. A drop of sap from a broken birch branch is also a symbol of a passing life.

Returning to the problems of ecology, it should be stated that a mythical image of a drop of moisture has the profound vital meaning. The emergence of life on Earth and the creation of conditions favorable for the existence of biota, including humans, is closely connected with a drop of forest moisture.

Everything is interconnected in nature; there is nothing superfluous. The presence of vital moisture on the continents of Earth is associated with forests. Russian scientists Gorshkov and Makarieva (2006) were the first to show one of the important functions forests perform in the biosphere. They called it the "biotic pump of atmospheric moisture." The essence of this mechanism is that the water evaporating from the surface of Earth condenses in the atmosphere and causes its rarefaction. Ascending air flow emerges and there is horizontal suction of air masses from the adjacent areas. This produces winds in the direction of the intense evaporation of moisture.

Before forests appeared, evaporation took place from the oceanic surface and was practically absent on land. Air was moving mainly toward the ocean and thus precipitation was developing only over the surface of the oceans. The land remained lacking moisture. The same situation is characteristic now of the desert regions.

For the birth of terrestrial life it was necessary that Earth was covered with vegetation, which is capable of evaporating more moisture than the oceanic surface. Forest cover was forming over billions of years of Earth's evolution. Forrest store approximately 85% of the terrestrial biomass. Forests with their branched crowns have a huge total surface of foliage and they evaporate more moisture than the oceanic surface. Forests thus "pump" water from the ocean and provide the land with moisture necessary for terrestrial ecosystems.

10.3   Green Economy

The so-called "green economy" has been widely advocated. This economic term has a purely symbolic character and does not assume any new strategies for the interaction of humans and nature. It is based on the well-known principles of human behavior in the natural environment. The green economy means a wide range of approaches of modern economy, focused on nature protection and rational nature management based on the use of modern resource-saving, nature-friendly, and adaptive technologies. The transition to such an economy is of fundamental importance from the point of view of the immediate implementation of these principles, especially considering the rate of modern degradation of the biosphere.

The very term "green economy" indicates that in the process of economic development, the most important priority is to preserve the green cover of the planet. Biota provides for the aesthetic appeal of landscapes, absorption of CO_2 and thereby reduction of the content of greenhouse gases in the atmosphere, and the supply of life-sustaining moisture, in the form of precipitation, to the continents. Without this, Earth would turn into a lifeless desert.

The most important reservoir of biodiversity and biospheric resources are forests—the lungs of the planet. Forests accumulate more than half of carbon on Earth. According to various sources, the area of forests is decreasing worldwide, as it is being transformed into industrial, agricultural, and urbanized landscapes. It is believed that the speed of forest area reduction in the world is 14.6 million ha/yr, and the recovery rate is 5.2 million ha/yr, i.e., forest area decreases annually by 9.4 million ha or by 0.26%.

In Russia, this situation is less acute because of the enormous forest areas. The forests of Russia make up 22% of world forests. About 69% of the territory of Russia remains covered with forests. A significant part of the currently existing forest tracts is still not affected by economic activity. In recent years, the area of forests in Russia has even slightly increased due to the decline in industrial timber production. However, a number of serious problems remain in the forestry sector, mainly due to the liquidation of state forest protection entities. Resultantly, the culture of forest management has declined—the extent of poaching, forest fires, forest diseases, uncontrolled timber trade, etc., has grown.

10.4 Ethical Principles and Nature

The dramatic environmental situation demands that mankind extends ethical moral principles in treating nature. Ethical principles mean equal consideration and responsibility toward nature on behalf of all people, regardless of social status, religion, ideology, level of education, age, etc. Ethical principles should protect nature from purely pragmatic human actions—technogenesis. All sensible and genuinely civil people should propagate ethical attitude toward nature. The most important value of life is the existence within the same emotional environment with nature, and not within the constructed pyramids of commercial satisfaction.

Ethical attitude toward nature is a genetic element of a human being. It is hard to find a person on Earth who does not like nature. But the genetic background of different people manifests itself in different ways and this is evident in human behavior when dealing with nature.

Immorality in relation to nature is especially striking when some entrepreneurs, in order to achieve commercial success, zombify people through poor-quality advertising by preying on their affinity for natural values.

The situation with drinking water is one of the examples. Tap water in many urban areas remains still quite suitable for drinking. Nevertheless, advertising intimidates residents, encourages the purchase of drinking water, often bottled by the advertiser from the same tap water with minimum cleaning, but labeled "drinking water" and often "mineral" or even "holy" water. It would be much more ethical if these entrepreneurs instead put treatment plants for the water supply systems of their cities.

Often a manufacturer of low-quality means of production polluting the environment also makes products for its clean-up. This is immoral in the same way as an arms manufacturer supplying weapons to both warring parties—they should be continuously at war to maintain a high demand for armaments.

11 The Future of Civilization. V. I. Vernadsky and the Noosphere

11.1 The History of Environmental Crises on Earth

The first representatives of the genus Homo, *Homo sapiens*, appeared about two million years ago in Africa. During the period from 1.7 to 0.7 million years ago, *Homo sapiens* migrated through the Middle East to Europe and Asia. The total number of people on the planet at that time did not exceed two million individuals.

The oldest representative of *Homo sapiens* in Europe were Cro-Magnons who lived there 40,000–10,000 years ago, which corresponds to the late Neolithic. Cro-Magnons were collectors and hunters; they possessed speech and used fire and primitive craft tools, which allowed them to stand-out from their ancestors.

The emergence of humans is the most important stage in the evolution of the biosphere. It allowed V. I. Vernadsky to state that a new geological force—Cro-Magnon, appeared on Earth; the influence of primitive humans on the environment had started. However, the scale of the impact was limited and determined solely by the muscular energy of humans. Therefore, changes in the biosphere did not exceed the threshold of stability and did not cause irreversible damage.

Despite this limited influence, the continuous evolution of the biosphere has led, over many of the millennia of the existence of human population, to several ecological crises. Therefore, the history of the biosphere is a chain of local and global crises with unpredictable consequences. The most known is the global crisis, called the "Neolithic revolution," caused by the extermination of large animals and other Pleistocene megafauna on Earth, which were food for humans. This crisis was resolved through a radical change in behavior and the transition to a new way of obtaining life-sustaining resources—from hunting wild animals to own food production. Tribes that could not follow this way died out and disappeared from the face of Earth. The number of Cro-Magnon people, as a result of this revolution, decreased by almost an order of magnitude, but they survived.

Thus, the Neolithic revolution was accompanied by the emergence of a new type of nature management. Such environmental crises are not instantaneous; any biological community has time for restructuring and, therefore, there are some chances of survival. This is true of the Neolithic revolution—tribes that did not switch to animal-husbandry and crop-farming disappeared, and those tribes that could change the way of obtaining life-sustaining resources survived.

The Neolithic revolution accelerated manyfold the transformation of civilization, creating fundamentally new incentives for its development.

11.2   A Brewing Ecological Crisis

People now live in the era of industrial, or scientific-technological, revolution. The Neolithic ecological crisis on Earth lasted about 10,000 years, while the era of the industrial revolution has lasted only 200 years. Despite such a short period, modern society is rapidly approaching a new global environmental crisis associated with the unprecedented technical means of modern society—a thousandfold increase in the impact on the environment, compared with the impact of primitive humans. It is believed that 30% of Earth is already in a degraded state and cannot provide natural resources for people living there.

Many environmental problems remain unresolved in the modern world; the most crucial problems are: (1) the continuing degradation of the biosphere as a result of exceedance of the global ecosystem ecological capacity threshold limits; (2) the growth of the overall pollution of the environment, including pollution by previously unknown chemical compounds; (3) global climate change and increase in the number of catastrophic natural phenomena; and (4) the deterioration of the human genome.

The concept of the modern human existence on Earth retains certain features of the paradigm of the Neolithic revolution—our society continues to consume (destroy) natural resources, thereby causing degradation of the biosphere. In addition to the depletion of life-sustaining resources, there is also chemical and physical pollution of the environment, which intensifies the degradation of the biosphere.

Modern scientific and technological progress is one-sided. It aims at promoting technogenesis and increasing the welfare of a small part of world inhabitants, while the global biospheric challenges remain unresolved. Unfortunately, the ruling elite of many countries diverts public attention away from this issue and is silent about the looming severe global ecological crisis caused by modern civilization, whose consumption exceeds the ecological capacity of the biosphere.

An important feature of the evolution of the biosphere is its ability to replace "used" biotic systems with new biological elements better adapted to changing parameters of the biosphere. If a certain biological species begins to dominate in the biosphere, i.e., becomes a monopolist, it does not have to compete, which leads to its rapid degradation. Such a monopolist reduces the ability to adapt and even a small change in the environment and external conditions leads to complete extinction. Humans may represent such a monopolist in the modern biosphere.

For millions of years, biota acted as a regulator of the functioning of the biosphere; it maintained the parameters of the biosphere in a narrow corridor, favorable for the existence of *Homo sapiens* as a biological species, despite powerful external influences (solar activity, volcanism, meteoric rain, geodynamic movement, etc.). Now the source of the main impacts on the biosphere are humans whose activity leads to severe deterioration of the biosphere. Will the biosphere survive this onslaught and be able to remain within the safe global environmental limits? If the biosphere is incapable of further controlling changes in its conditions and exceeds threshold limits, many living organisms, including humans, obviously would not be able to adapt to these new conditions and would inevitably perish. As a result, the biosphere will move to another "orbit" of its development, but without humans.

The emerging environmental crisis demands radical change in the strategy of technogenesis in order to observe the external and internal relationships and laws of the biosphere. However, humans are increasingly separating themselves from the biosphere, seeking to rise above it, and completely replace it with the technosphere. This violates one of V. I. Verdandsky's precepts, *"A man is inseparable from the biosphere and can exist only in the biosphere."*

There remains the only one option. Humans can only rely on Collective Intelligence and use it to rebuild their ecological niche. It remains uncertain whether there is sufficient time to address this. Unfortunately, the rate of destruction of the ecological niche of Humans now exceeds the rate of maturation of Common Intelligence, which quickly brings us closer to a new ecological catastrophe. The problem is further aggravated by the fact that most people are oblivious to this issue and do not bother to ponder the questions about the continued existence of life on Earth.

The structure of a rationally organized society in relation to nature is not yet quite clear, especially considering the fact that in different regions of the world, this structure may also vary. The only thing quite clear is that society must follow the collectively worked out requirements that ensure equality of rights for all citizens in relation of nature and its resources.

11.3 Will Civilization Survive?

In order to avoid the impending ecological catastrophe, it is necessary to mobilize the entire potential of scientific and technological progress and to focus it on the solution of the vital issue of finding ways to further the evolution of the biosphere with the goal of preserving human civilization in completely different conditions, termed the "noosphere" by Vernadsky (1944, 1989).

The *Noosphere* was described by V. I. Vernadsky as part of the biosphere and the natural stage of its development, organized by civilization. The term "noosphere" was proposed earlier by the French scientists P. T. de Chardin and E. Le Roy, though V. I. Vernadsky gave a fundamentally different interpretation of this concept; he proposed treating this notion from the geo-centric or nature-centric point of view instead of its anthropo-centric interpretation. V. I. Vernadsky emphasized two points: (1) humans do not control the development of the noosphere, but only participate in its evolution; (2) the noosphere is not the sphere of technology, not the sphere of humans and society, but the sphere of Intelligence—the "thinking shell" of the planet.

The name of V. I. Vernadsky is closely associated with the scientific content of this term. He believed that humans became an important geological force that transforms our planet. To survive on Earth, people must take responsibility not only for the development and preservation of society, but also for the biosphere as a whole. As a result of such interaction, the biosphere will transition to a new state—the noosphere. The co-evolution of humans and the biosphere will obviously be the main theoretical issue of scientific research focused on studies of the conditions of human existence on the planet.

V. I. Vernadsky considered the noosphere to be the sphere of the evolution of intelligence, realized through scientific thought, which gradually becomes a "planetary phenomenon" (Vernadsky 2000). He used the term "scientific thought" to define the organized, commutative, collective thought of all people, capable of influencing the course of biospheric processes. The newly created geological factor—scientific thought, changes the phenomena of life, geological processes, and the energy of the planet. This side of the evolution of human scientific thought represents a natural phenomenon. It was generated in the course of the evolution of the biosphere and is naturally changing with time. However, scientific thought does not exist apart from the human mind, it is the product of human creative energy. Thus, the noosphere, in V. I. Vernadsky's understanding, is a regular stage in the development of the biosphere, organized by civilization, which began with the emergence of human conciseness, continues in modern time, and will continue in the future.

The concept of the noosphere demands humans to find a new form of existence allowing them to survive in the biosphere. The main problem is to provide society with life-supporting resources and, first of all, with nutrition. One possible way of solving this problem is the implementation of a new biospheric revolution (like the Neolithic one) by synthesizing nutrition (including proteins) using solar energy and inert components of the natural environment (bypassing the consumption of animal and plant foods).

The concept of the existence of humans as a planetary entity which consumes extraterrestrial energy is purely hypothetical with many unknowns. There will be a serious discussion about this issue and it is important to begin such a discussion to understand better the role and place of humans in the biosphere and the conditions for our survival. One of the important moments of this discussion is the realization that the basis of life and development of society is not the economy, but the laws of the biosphere.

In 1925, in his lecture "Autotrophic Humanity," at Paris-Sorbonne University, V. I. Vernadsky expressed another prognostic idea of ensuring the human existence through the use of technology for obtaining synthetic food based on the return of waste to the biogeochemical cycles of the biosphere.

The notion of autotrophy was expressed by V. I. Vernadsky following the success in the field of inorganic synthesis of hydrocarbons, proteins, and fats. He believed that by deepening these studies, it will be possible to replace traditional biogenic food products with artificial ones. However, the subsequent studies have shown that organic food differs from artificial by different ratios of isotopes entering into these compounds. Therefore, it becomes necessary to set up fundamental research on the essence and technology of food synthesis.

Thus, the path in the epoch of the noosphere seems to be, to a large extent, still unknown and thorny. Nevertheless, the solution to this problem cannot be postponed. The main question of whether the transition to the epoch of the noosphere is possible in principle, remains unanswered. If it proves possible, there may be a real revolution in the fate of humans and, most importantly, there will be conditions for the continued existence of civilization.

Finalizing the discussion on the noosphere, it becomes clear that the world around us and life on Earth are complex, diverse, and not yet completely understood. At the root of the notion of the noosphere is V. I. Vernadsky—a thinker of the planetary caliber. In his reflections on the future of human civilization, V. I. Vernadsky was pushing the boundaries of traditional scientific knowledge and approaches. He wrote that scientific creative thought transcends logic. At the same time, he realized that the understanding of the enormity of the problem of interaction between society and nature is still at the initial stage. V. I. Vernadsky wrote, "The reign of my ideas is yet to come" (Fig. 26).

12   Conclusion

Human civilization is drawn into the global ecological crisis, completely unlike the past critical situations in the history of Earth as a geological entity. The coming crisis makes the modern world even more dangerous, since it creates a threat of not only military but of "peaceful" destruction of human population by the evolving ecological catastrophe. Obviously, this is an even more detrimental and painful process than a nuclear conflict.

Understanding that the biosphere is rapidly degrading on a large part of our planet has not become a phenomenon of public consciousness. It is quite obvious that the impending ecological crisis cannot be prevented with purely technical and administrative measures. Widespread use of more advanced technologies in human economic activities and stricter measures of the natural resources rational use can only delay the onset of the crisis, ease the consequences, and give people timeout for finding more radical solutions that have yet to be developed by studying the laws of development of the biosphere and the ideas of V. I. Vernadsky about the noosphere. It is crucial to recognize the new stage in human civilization existence in the biosphere because human civilization has exhausted its developmental potential in its present form.

This conclusion leads to the idea that the development of technogenesis and environmental degradation can lead to the cessation of existence of a biological species

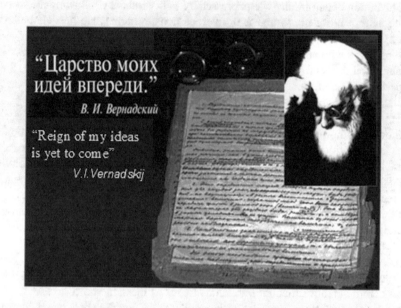

Fig. 26 V. I. Vernadsky is a thinker of the planetary caliber

Homo sapiens. Civilization, obviously, can still exist for quite some time by allowing people to migrate to less polluted territories. But such massive migration can lead to aggravation of international relations, which has been already the case. It is unlikely that any nation will be willing to give away their territories to aliens from other countries. Therefore, armed confrontations are inevitable, which in the age of nuclear weapons can lead to the final global catastrophe.

Some sources propagate vigorously the ideas of the extinction of a large part of world population and the preservation of a limited number of humans on the planet. This is the so-called theory of the "golden billion," according to which the population of about one billion can remain on Earth in conditions of complete regeneration of life-sustaining resources. However, the laws of evolution of the biosphere are not taken into account, but these laws dictate that: (a) a species that has entered the path of degradation eventually dies out completely and irreversibly and (b) it is impossible to restore the extinct or degraded biological species in the biosphere.

The question arises of whether there are chances of survival and preservation in such a situation. In the opinion of Moiseyev (1999, 2007), the future of human civilization can be realized only through a compromise that must be achieved between human society and nature. Such processes, in one form or another, have always existed in the living world for two million years of the *Homo sapiens* history, which led each time to the mutual adaptation of humans and the hosting biosphere. The question is: what is the course of action now?

This means that it is necessary to abandon the old stereotypes and move to the positions of a new ideology that relies on the biosphere-oriented worldview rather than depends solely on economic and social constructions.

The essence of the biosphere-centric worldview is the fact that humanity is only part of the biosphere and of its main component—biota. The biosphere is incommensurably more complex than human civilization. In many ways it is incomprehensible to the human mind and, therefore, humans cannot substitute it with a technogenic system.

The second important aspect of biosphere-centric philosophy is the rejection of the super-ideology of modernism, unfoundedly putting humans and their work to the level of divine power and thereby misleading humanity as to our possible existence in this world. Humanity cannot be held higher than the biosphere that gave rise to us, providing us with nutrition and creating the optimal environment for the existence.

To implement the biosphere-centric ideas, specific scenarios should be developed as soon as possible before it is too late. Scientists, political leaders, professionals, mass media, etc.—all those who can clearly perceive the future, should play the most important role as has been the case in the past.

Politicians who dream of the domination of their countries in the world must urgently change their rhetoric, for above all, a more abysmal threat is brewing—the environmental crisis. Therefore, humans on Earth have a common goal—to preserve the human race and to avoid the end of civilization.

References

Agenda for the twenty-first century, in *UN International Conference on the Environment*, Rio de Janeiro (1992)

V.I. Bogoyavlensky, The threat of catastrophic gas emissions from the Arctic permafrost zone. Funnels of Yamal and Taymyr. Drill. Oil **10**, 4–8 (2014)

M.I. Budyko, *Evolution of the Biosphere* (Hydrometeoizdat, Moscow, 1984), p. 488

Centre for Research on the Epidemiology of Disasters. (http://www.emdat.be/)

Delovaya Gazeta (Business Daily). Guide to international business, 120/121. October–November, 2016

R.L. Folk, Nanobacteria in the natural environmental and in medicine. Microbiology **7**, 87–89 (1998)

V.G. Gorshkov, Structure of biospheric energy flows. Bot. J. **65**(11), 1579–1590 (1980)

V.G. Gorshkov, A.M. Makarieva, The biotic pump of atmospheric moisture as the driver of the hydrological cycle on land. Hydrogeol. Earth Syst. Sci. 2621–2673 (2006)

V.G. Gorshkov, A.M. Makarieva, Dependence of heterozygosity on the body weight of mammals. Rep. Russ. Acad. Sci. **335**(3), 418–421 (1997)

S.K. Gulyov, V.M. Kattsev, O.N. Solomina, Global warming continues. Bull. Russ. Acad. Sci. **78**(1), 20–27 (2008)

IPCC Fifth Assessment Report, http://ipcc.ch/report/ar5/index.shtml (2014)

IPCC Fourth Assessment Report, http://ipcc.ch/report/ar4/ (2007)

S.P. Kapitsa, The global problem of humanity. Bull. Russ. Acad. Sci. **68**(3), 234–241 (1998)

V.M. Kotlyakov, C. Lorius, Four climate cycles according to the ice core data from deep drilling at the Vostok station in Antarctica, in *Proceedings of the Russian Academy of Sciences. Ser. Geographical*, vol. 1 (2000), pp. 18–19

V.M. Kotlyakov, History of Earth's climate based on the data of deep drilling in Antarctica. Priroda (Nature) **5**, 3–9 (2012a)

V.M. Kotlyakov, On the causes and consequences of modern climate change. Sun Earth Phys. **21**(134), 110–114 (2012b)

V.I. Larionov, N.I. Frolova, *General Methodology of Risk Assessment. Encyclopedia of Safety: Construction, Industry, Ecology*, vol. 1 (Nauka, 2005), pp. 2–34

N.N. Moiseyev, in *To Be or Not to Be for Humanity* (Ulyanovsk Publishing House, Moscow, 1999), 288 p

N.N. Moiseyev, Modern natural science and problems of interaction between nature and society. Ecol. Life **8**(69), 10–14 (2007)

I.I. Mokhov, A.V. Yeliseyev, Modeling of global climate changes in the XX–XXIII centuries under scenarios of the anthropogenic impact of RCP. DAN **443**(6), 732–736 (2012)

G.B. Naumov, in *Geochemistry of the Biosphere: A Manual for Students of Institutions of Higher Education* (Publishing Center "Academy", Moscow, 2010), p. 384

A.I. Nikitin, in *Harmful Factors and the Human Reproductive System*. Responsibility to future generations. SPb.: ELBI-SPB. (2005), 264 p

V.I. Osipov, N.I. Frolova, S.P. Suschev, V.I. Larionov, in *Assessment of Seismic and Natural Risks for the Population and Territory of the Russian Federation. Extreme Events and Catastrophes*, vol. 2 (Publishing house "Probel 2000", Moscow, 2011), pp. 28–48

V.I. Osipov, S.P. Suschev, N.I. Frolova, A.N. Ugarov, S.V. Kozharinov, T.V. Barskaya, Assessment of the seismic risk of the territory of Great Sochi. Geoecology **1**, 5–21 (2015)

B.N. Porfiryev, Economic consequences of catastrophic flooding in the Far East in 2013. Bull. Russ. Acad. Sci. **2**, 30–39 (2015)

M.M. Prishvin, *Green Sound* (Pravda, Moscow, 1983), 480 p

V.S. Shchukin, The transition of man from *Homo sapiens* to *Homo immortalis* as an indispensable condition for the transformation of the biosphere into the noosphere. Evolution **5–6**(12), 94–95 (2010)

V.N. Sukachev, Biogeocenosis as an expression of the interaction of living and inanimate nature on Earth's surface: the correlation of the concepts "biogeocenosis," "ecosystem," "geographical landscape," and "facies", in *Fundamentals of Forest Biogeocenology*, ed. by V.N. Sukachev, N.V. Dylis (Nauka, Moscow, 1964), pp. 5–49

The second assessment report of Roshydromet on climate change and its consequences on the territory of the Russian Federation (Roshydromet, Moscow, 2014), 93 p

The state report, in *On the State and on the Protection of the Environment in 2013* (Roshydromet, 2014)

N.V. Vakulenko, V.M. Kotlyakov, A.S. Monich, D.M. Sonechkin, Symmetry of the glacial cycles of the Late Pleistocene according to the data of the Vostok station and "Dome C" in Antarctica. Far East Branch Russ. Acad. Sci. **407**(1), 111–114 (2006)

V.I. Vernadsky, A few words about the noosphere. Adv. Mod. Biol. **18**(2), 118–120 (1944)

V.I. Vernadsky, in *Biosphere and Noosphere* (Nauka, Moscow, 1989), 261 p

V.I. Vernadsky, Scientific thought as a planetary phenomenon, in *Proceedings on the Philosophy of Natural Science* (Nauka, Moscow, 2000), pp. 316–451

A.N. Vinogradov, Y.A. Vinogradov, S.V. Baranov et al., The risk factors associated with the degradation of the cryolithosphere in the Western Arctic and the problems of their geophysical monitoring, in *Collection of Scientific Works*, ed. by V. Pavlenko (Arkhangelsk, 2016), pp. 64–70

V.T. Volkov, N.N. Volkova, G.V. Smirnov, in *Biomineralization in the Human Body and Animals* (Tandem-Art, Tomsk, 2004), p. 500

World, in *World Population Prospects: The 2012 Revision, Highlights and Advance Tables* (UN DESA, Population division, Working Paper, 2013. No. ESA/P/WP. 227 (Electronic Recourses)). https://esa.un.org/unpd/wpp/publications/Files/WPP2012_HIGHLIGHTS.pdf. Retrieved 11 Feb 2017

A.V. Yablokov, V.F. Levchenko, A.S. Kerzhentsev, Essays on biosphere. 1. There is way out: a transition to a controlled evolution of the biosphere. Philos. Cosmol. **14**, 92–118 (2015)

Printed in the United States
By Bookmasters